Smart Energy

Solar Energy

Smart Energy
From Fire-Making to the Post-Carbon World

Jianping Liu, Shaoqiang Chen, and Tao Liu

CRC Press
Taylor & Francis Group
Boca Raton London New York

CRC Press is an imprint of the
Taylor & Francis Group, an **informa** business

CRC Press
Taylor & Francis Group
6000 Broken Sound Parkway NW, Suite 300
Boca Raton, FL 33487-2742

First issued in paperback 2020

© 2017 by Taylor & Francis Group, LLC
CRC Press is an imprint of Taylor & Francis Group, an Informa business

No claim to original U.S. Government works

ISBN 13: 978-0-367-57356-0 (pbk)
ISBN 13: 978-1-4987-7648-6 (hbk)

Library of Congress Cataloging-in-Publication Data

Names: Liu, Jianping, 1965- author. | Chen, Shaoqiang, author. | Liu, Tao (Environmental engineer), author.
Title: Smart energy : from fire-making to the post-carbon world / Jianping Liu, Shaoqiang Chen, Tao Liu.
Description: Boca Raton : Taylor & Francis, a CRC title, part of the Taylor & Francis imprint, a member of the Taylor & Francis Group, the academic division of T&F Informa, plc, [2017] | Includes bibliographical references.
Identifiers: LCCN 2016058863 | ISBN 9781498776486 (hb : alk. paper)
Subjects: LCSH: Power resources--History. | Renewable energy sources. | Energy policy.--China.
Classification: LCC TJ163.2 L587 2017 | DDC 333.7909--dc23
LC record available at https://lccn.loc.gov/2016058863

Visit the Taylor & Francis Web site at
http://www.taylorandfrancis.com

and the CRC Press Web site at
http://www.crcpress.com

Contents

Foreword

Silent Spring, written by Rachel Carson in the 1950s, attracted the public's attention to the problem of the environment. In the 1980s, the United Nations officially put forward the concept of sustainable development that was valued by countries all over the world and has since become a critical problem related to human beings' survival. After several thousands of years of agricultural civilization and hundred years of industrial civilization, human beings are entering the knowledge society in which there is a need for a new civilization—the ecological civilization of saving resources, protecting ecology, and living in harmony with nature. The people-oriented, comprehensive, coordinated, and sustainable scientific outlook on development proposed by the Communist Party of China (CPC) Central Committee is a powerful response to such a civilization. The 18th CPC National Congress called for the building of an ecological civilization and a beautiful China, which is certain to further incite the whole world to attach greater importance to energy and environment problems.

If capital is the blood of economy, then energy is the food for economy. Both capital and energy are indispensable for the economic growth and social development of a country. For a long period of time, mainly relying on the consumption of coal, petroleum, natural gas, and other fossil energies, human beings have achieved economic and social development. However, fossil energy is nonrenewable and will run out eventually. Besides, global climate change caused by enormous carbon dioxide emissions that is mainly attributable to fossil fuel use has become the biggest threat to human beings. Therefore, new energy has become a hot topic and has gained a lot of attention from many countries and governments in recent years.

At present, there are many different definitions of new energy, such as "green energy," "renewable energy," "low carbon energy," etc., whose implications are not completely the same. In my opinion, new energy can be defined at two levels. In a narrow sense, it includes wind energy, solar energy, tidal energy, geothermal energy, biofuel, and other energy forms that were not widely used before. In a broad sense, in addition to the aforementioned, it should also contain nuclear energy, hydropower energy, and even clean coal technology.

After the global financial crisis of recent years, every country in the world has to look for new economic growth points and adopt a new development pattern. I think new energy is certain to be a new economic growth point. From a wider historical perspective, the first three Industrial Revolutions mankind experienced were led by the steam engine, electric power, and the computer, respectively. Each of them brought the industrial development level to a new height and benefited consumers. Therefore, several years ago, I had stated that the fourth Industrial Revolution will be an energy revolution driven by new energy.

At present, it is the western countries' view of the future that dominates global public opinion. We as Chinese should also have our own voices and opinions affecting the world in the 21st century. It is the need of the hour to present important and systematic opinions about the issue of energy as this has a significant effect on the world. In my opinion, the formulation and systematic thinking of energy revolution integrating new energy, environment, economic development, social progress, China, and the world as a whole is of great significance.

Based on this, when three young and middle-aged authors who were, respectively, born in the 1960s, 1970s, and 1980s asked me, through my secretary, to write the preface for their new book *Smart Energy—From Fire-making to the Post-Carbon World*, although I had never met them before and was quite busy with other work, I agreed to do it after I read the book.

The "smart energy" the authors talk about in this book is defined as "the energy form which integrates human beings' intelligence in the whole process of energy development, use, production and consumption," "and which has self-organization, self-check, self-balance, self-optimization and other human brain functions to satisfy the requirements of system, safety, cleanness and economy." Starting from a macrohistory perspective, the authors discuss energy and civilization, which is a topic of common concern, and explore smart energy and human development. They link theory with practice as much as they can, combine contentions and arguments, go from the easy to the difficult, and excel with regard to illustrations and text, which brings in a new style of writing—the "quasi-academic book."

The authors put forward some interesting viewpoints, for example, smart energy takes energy form as a carrier, but is not limited to it, and it combines energy form, energy technology, and energy system; smart energy embodies human intelligence that supports the advance of human civilization; the developing ecological civilization and any upgraded civilization in the future cannot exist without the support of smart energy; different forms of human civilization have different requirements of energy, and the upgraded civilization in the future must require energy of higher intelligence, etc. Although it is only statements of one school of thought, a Chinese saying goes "the benevolent see benevolence, the wise see wisdom." The systematic thinking of energy is indeed a useful trial.

Finally, I would like to point out that the development trend of new energy in China is satisfying in general, but faces serious challenges. The challenges include: (1) lack of a considerate and careful plan for the development of new energy, at present, and many controversies over the development of new energy; (2) less key technologies and innovation in new energy; (3) universally higher cost of new energy at present; and (4) the high dependency on fossil energy, currently as high as 90% in China, which is difficult to change in a short period, although the public are eager to resolve haze and other serious environment problems as soon as possible. Therefore, the development of new energy in China must be on a firm footing and must move forward with calm consideration to avoid excessive hype. The coordinated development of energy and environment in China must be promoted from economic, technological, and political aspects, with firm direction and even steps.

I hope the publication of this book will catch the attention and arouse the thoughts of people with insights who could make due contributions to resolving the serious problems of energy and environment that curb the economic development of our country in the future.

Cheng Siwei
*Vice President of the Standing Committee
of the Tenth National People's Congress,
China*

Preface

Innumerable years have passed since mankind came into being in this world. In the endless long river of history, we have been ceaselessly familiar with, adapted to, and aware of nature and have gradually found and utilized fire, thus, beginning to obtain the magic weapon—energy—with which we could change heaven and earth and by which we have achieved myriad successes and dreams. Our common fate has continuously changed and improved. Especially now, in less than ten thousand years since the time of written records, energy has been a major driving force of human civilization and is a link that removes the distances created by the seas and unites the continents in the world as a whole.

Inquiry into the Origin of Life

Where did we come from? Where do we go? These are the ultimate and unanswerable questions that we, looking up at the stars, have pondered upon since time immemorial.

The earth revolves round the sun, one of over 10 billion stars in a galaxy that is one of several million galaxies in the universe. The universe, the sun, and the earth were, respectively, formed about 15 billion years, 5 billion years, and 4.5 billion years ago, and this was calculated according to the theory of the Big Bang. Regarding the origin of life, there are not only religious and scientific explanations, but even within the scientific community, controversy exists between proponents of earth autogeny and outside help. The only concrete conclusion that we can arrive at is that by some miracle or chance there is life on the earth, which has continuously evolved from low level to high level, forming diversified and colorful biospheres.

If the appearance of life on earth is a miracle, the appearance of human beings is a myth and a mystery. There have been different legends about the origin of human beings since time immemorial, but no convincing explanation so far. Although no final conclusion has yet been reached on this matter, Darwin's theory of evolution, which says humankind evolved from apes, has been accepted by most people. At first, a kind of anthropoid evolved from the lower apes. Later, about six million years ago, "humans and apes bowed to each other and said goodbye,"* and each went their own way. From then on, mankind has made their mark on the colorful stage that is the earth.

Probing the Path of Civilization

We live on the earth with many different living things. As time has brought great changes to the world, mankind has also experienced difficult changes from being a *Homo habilis*, ape man, and ancient man to a new man, and from ignorance and savagery to civilization.

"Only a few stones were polished in the childhood."† The rise of the human civilization went through a preparation period that was nearly four million years long. In the Old Stone Age, our ancestors invented stone implement-making technology, fire-making technology, and language, which are, respectively, seen as the embryo and origin of machine-making technology, energy conversion technology, and the present information technology. The invention and evolution of those three technologies laid the foundation for the origin and development of the human civilization. The use of fire is of extremely special significance for mankind as it helped to master energy, the magic weapon: "Making fire by friction made people dominate a kind of natural force for the first time and differed them from animals in the end"‡ Why do we say that the human being is the wisest of all creatures? Because it is humans who created tools and dominated fire. At that accidental and inevitable moment when our ancestors groped and knocked two flint stones together or drilled wood to make fire, humans accepted nature's selection and became the most intelligent of all creatures.

The important turning point in the course of the generation of human civilization was the New Stone Age, and agriculture became its accelerator. Because the growth of crops were periodic, humankind who lived in areas suitable for farming settled down from nomadism and increased labor productivity greatly. With surplus products, commodity exchange, private property, and private ownership came into being. As social products could support more people, division of labor, different social groups, and various interest groups also came into being. With disputes caused by interest distribution, military democracy arose. Because of more frequent wars under military democracy, military leaders and gentile aristocracies gradually separated from equal clan members, rose to power, and built corresponding powerful ruling organizations; thus, countries were born. The early human civilization, like the first rays of the morning sun, first shone on the region from 20° E to 120° E longitude and from 20° N to 40° N latitude,

---

* *Hexinlang, Reading History* by Mao Zedong.
† *Hexinlang, Reading History* by Mao Zedong.
‡ *Selected Works of Karl Marx and Friedrich Engels.* Vol. 3. People's Publishing House, 1995.

covering Mesopotamia, the Nile Valley, Yangtze River Valley and Yellow River Valley, Indus Valley, and the Aegean Sea coast.

Casting a Brilliant Foundation

Mankind entered the civilization of recorded history six thousand years ago. The history before this is generally called prehistory, with millennium as units. The unit of the history of civilization continues to be shortened, gradually becoming one hundred years and even ten years. Like the birth of mankind and the origin of life, the history of civilization is very short, compared with prehistory. Having entered civilization, humans from different areas did not develop from the same starting line or at the same speed but independently and separately until after AD 1500 when the west started the Age of Exploration by sailing on oceans first with sailboats and then with steamships, connecting the world and making it an inseparable whole. From then on, human civilization began to bid farewell to the agricultural civilization, stepped into the industrial civilization, and then gradually transited to the new era of information civilization.

Important scientific discoveries and inventions have continuously sprung up and thoroughly changed human being's production mode and lifestyle since the industrial civilization. The British invented the steam engine in the late half of the 18th century, burning coal instead of wood, which was a great change in energy form. Later, the invention of the internal combustion engine let petroleum replace coal, which has increased productive efficiency greatly. The currently popular airplanes and automobiles are all masterpieces that cannot work without the internal combustion engine. The invention of the generator by Siemens is of epoch-making significance, similar to the invention of the steam engine by Watt. Electric energy provides huge amounts of power for human civilization. All kinds of modern communication tools continuously spring up, creating the brilliant modern information civilization.

Starting the Future Engine

The use of coal, petroleum, and electricity has brought us unprecedented economic prosperity, while making the modern society more and more seriously dependent on fossil energy. For a century, the excessive exploitation and use of fossil energy has generated a series of severe problems: Global warming, extreme climate, environmental disruption, drying up of rivers, resource shortage, disputes about energy getting worse, etc.

Solving the abovementioned problems depends on scientific innovation and breakthrough as well as the fundamental change of the mode of production, consumption, and even social system. This is a severe challenge and arduous mission.

Everyone living on the earth should shoulder this unavoidable responsibility. We should join hands and work together to realize the existence and development of human civilization, which is of common interest to all human beings.

Looking back at the past ten thousand years of history that is filled with glories and dreams, and wonders and brilliance, we are filled with ambition; looking back at the past ten thousand years of our history that is also glutted with hardships and sufferings, and setbacks and sorrows, we have a thousand misgivings. Over the past ten thousand years, our ancestors discovered fire when they were ignorant; with fire, an impassable gulf between mankind and other creatures was formed, and mankind became the wisest of all creatures; our ancestors used cattle and horses to plough and employed water wheels and windmills for irrigation and processing, by which they lived and evolved; they then utilized coal and steam engines to enter the great epoch of industrial revolution, striding forward, singing songs, never stopping from advancing; they used dazzling electricity to create the information civilization, which is progressing with each passing day.

In all these eras, energy was of prime importance. It was the elf that brought light; it was the warm and driving force of development and advancement; it was also the ghost that brought suffering, pain, stagnation, ruin, and horror. It played an irreplaceable role in each scene of civilization, pushing our history forward. Energy and civilization coexist and supplement each other. As human wisdom ceaselessly accumulates and develops, energy is constantly transformed and replaced. Firewood, animal power, wind power, water power, coal, petroleum, and electricity have provided us with a steady flow of driving force on our way forward. We have experienced the hunting–gathering civilization, agricultural civilization, industrial civilization, and information civilization. What will tomorrow's energy form, which will start the engine of the future and embrace the brilliant ecological civilization for us, be like?

About the Authors

Jianping Liu received his master's degree of Labor Economics, PhD in Industrial Economics from Renmin University of China, and his post-doctoral in Environmental Science and Engineering from Chinese Academy of Sciences. He then worked in the Finance Ministry, State Administration for Industry and Commerce, and the State Electricity Regulatory Commission. At present, he is division director of the Department of Development and Planning, National Energy Administration; research fellow of the Institute of Urban Environment, Chinese Academy of Sciences; and professor of China–EU Institute for Clean and Renewable Energy, Huazhong University of Science and Technology. He has published more than 20 papers on a wide range of topics in system reform and energy development. He has also written four books since 2006, and his latest book *Smart Energy—From Fire-making to the Post-Carbon World* (written in Chinese) won the Popular Science Books Award of the Chinese Society for Electrical Engineering and China's Weekly Reading Top 100 Books Award in 2013. The book was also nominated for the Popular Science Books Award by the China Science Writers' Association.

Shaoqiang Chen obtained a master's degree in finance from Zhongnan University of Economics and Law, a master's degree in public policy management from Hitotsubashi University, and a doctorate degree in economics from the Research Institute for Fiscal Science, Ministry of Finance. He now works in the Research Institute for Fiscal Science, Ministry of Finance, as researcher and supervisor of postgraduate students. His research interests lie in the field of fiscal and monetary policy, state-owned asset management, social security, and energy-saving and emission-reduction policies. He has already published dozens of articles in these areas.

Tao Liu obtained a master's degree in management from Changsha University of Science and a doctorate degree in economics from the Research Center for Eco-Environmental Sciences, Chinese Academy of Sciences. He is a postdoctoral scholar of the Institute of Economic Management, Tsinghua University.

Chapter 1

Approaching Energy: Our Driving Force of Development

Since ancient times, energy has always been the driving force for our life. Modern society cannot develop without coal, petroleum, natural gas, and electricity. Therefore, energy as a concept is quite commonly known. However, even though energy is useful, we get into trouble as we increase our conveniences. Let's look at the world of energy, amount of energy, and sources of energy and see why they are different songs from the same source, what kind of face-changing skill they have, and how they coexist with us in our life like a shadow following the body.

1.1 Different Songs with the Same Source: Energy, Amount of Energy, and Sources of Energy

1.1.1 Energy

Generally speaking, energy is the ability and capacity of physical systems to do work.* In physics, if an object can do work, and work on another, it is said to have energy or have the ability to do work. Energy is a dynamic concept. The different

___
* Work is referred to as the activity in which energy is changed from one form to another. It is an important concept in physics. When a force is applied to an object and makes it move through a distance along the direction of the force, the force is said to have done work.

forms of movement of matter correspond to different forms of energy, and energy of different forms can be converted from one form to another.

There is controversy regarding the classification of energy, but it can be roughly classified into radiant energy, mechanical energy, chemical energy, molecular energy, electromagnetic energy, and atomic energy based on the history of mankind's development and use of energy. With increased understanding and development of science and technology, more forms of energy can be identified, and so the classification of energy could be changed.

Radiant energy was the earliest form of energy used by mankind, and solar radiation is one of the most common and important sources of radiant energy (see Figure 1.1). We have been using it even before we fully understood it. Most of the creatures on earth receive solar radiation. Energy from solar radiation has been accumulating in different kinds of life-forms for a long time, just waiting for people to exploit it and use it for future.

Mechanical energy is the generic name given to the sum of kinetic energy and potential energy, and is one of the forms of energy that people knew about earlier and related to mechanical motion and relative position of the whole object. Kinetic energy is the energy the object produces when it moves, and the amount of energy depends on mass and speed. Potential energy is the energy required by the object to do any work, and it includes gravitational potential energy and elastic potential energy. The energy the object has when it is raised to a certain level above the ground is called gravitational potential energy, which is determined by height, mass, and gravity coefficient. The energy the object has when it undergoes elastic deformation is called elastic potential

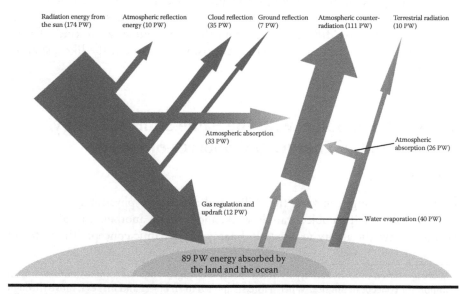

Figure 1.1　About half the incoming solar energy reaches the Earth's surface.

energy, which is determined by elastic coefficient and deformation. Mechanical energy is used in many aspects of life. All hydropower stations use the potential and kinetic energies of water to drive generators to generate electricity (see Figure 1.2). Additionally, raising a heavy hammer pile uses gravitational potential energy, while shooting arrows uses elastic potential energy.

Chemical energy is the energy that the object releases through chemical reaction, and this energy lies latent in different forms of matter. It cannot directly do work but can only be released when chemical change occurs and converts this energy to other forms. The energy released by combustion of a substance (see Figure 1.3), explosion of dynamite, respiration and photosynthesis of plants, chemical changes of food under the action of different digestive enzymes in the body, etc., are different forms of chemical energy.

Molecular energy refers to the sum of the kinetic energy produced by all the molecules inside an object when they are in irregular motion and the potential energy produced by the mutual action of the molecules (see Figure 1.4), also called internal energy. Different from mechanical energy, the quantity of the internal energy of the object is related to the heat generated by the motion of the molecules inside the object and their mutual action, at the micro level.

The kinetic energy produced by molecular heat motion is also called heat energy. The higher the object temperature, the higher the speed of molecular heat motion and the greater the heat energy. The heat energy can replace molecular energy (internal energy) in some cases.

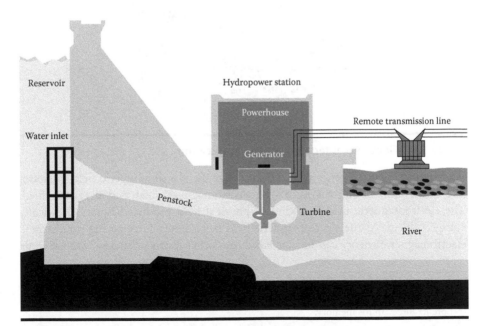

Figure 1.2   Hydropower is one important use of mechanical energy.

Figure 1.3 **Combustion is a kind of violent chemical reaction and produces much heat. (Photo by Lu Jinyang.)**

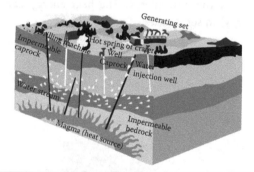

Figure 1.4 **Geothermal technology and application.**

Electromagnetic energy is the energy that the electromagnetic field possesses and is the sum of the energy of the electromagnetic field and magnetic field. The electromagnetic field is a kind of physical field produced by an electrically charged object. As the electromagnetic field has Lorentz force acting on charges*, electromagnetic energy can be converted into other forms of energy that are more

---

* Netherland physicist Lorentz (1853–1928) first raised the viewpoint that moving charges produced magnetic field, which in turn acts on the moving charges. This kind of force is called "Lorentz force" in his memory.

convenient for use, through doing work on the moving charges. Electric energy is the main form of electromagnetic energy. In daily life, hydropower generation, thermal power generation, and wind power generation all use this principle to produce electric energy using the various metal coils inside the generator and the relative motion of the electromagnetic field (see Figure 1.5).

Atomic energy is the energy that the neutron and proton in nucleus release when they are redistributed; it is also called nuclear energy and includes nuclear fission energy and nuclear fusion energy. Before atomic energy was identified, people only knew of chemical energy, for example, the kinetic energy produced by a car in motion, the thermal energy released by alcohol when it burns and is converted into carbon dioxide and water, and the electric energy of current giving out heat and light when it passes through electric stove wire. The release of these energies cannot change the mass of matter, but it can change energy from one form to another. In the early 20th century, scientists found that while uranium-235 nucleus was undergoing fission, it released 2–3 neutrons and a large amount of energy, much more than the energy released by chemical reaction— this was nuclear fission energy. The atomic bomb uses the energy released on nuclear fission to cause damage (see Figure 1.6). A nuclear power station also uses this principle to generate energy under safe and controlled conditions. The hydrogen bomb uses the principle of nuclear fusion; its power is much greater than that of the atomic bomb.

Figure 1.5 Shanghai magnetic suspension model operation line, which is driven by suspension force (namely the attraction and repulsion of magnets).

Figure 1.6   The mushroom cloud produced by the explosion of the atomic bomb "Little Boy" dropped on Hiroshima, Japan, on August 6, 1945.

Don't Belittle "Water Drops"

Water, having a chemical formula H_2O, is an inorganic compound consisting of two elements, hydrogen and oxygen. It is a colorless, odorless, and transparent liquid at ordinary temperature and pressure, can be seen everywhere in our life, and even makes up 70% of our body's weight. Don't belittle a drop of water—it has radiant energy, chemical energy, mechanical energy, molecular energy, electromagnetic energy, and atomic energy. If you don't believe this, let's take a trip with nature.

A piece of ice that has remained dormant for thousands of years in the Himalayas is crystalline and translucent. When strong winds blow away the snow covering her, she opens her eyes and stretches. She turns over by the virtue of wind and begins to roll faster and faster, and can't stop. The piece of ice on the Himalayas that is thousands of meters above sea level has an inbuilt powerful mechanical energy—gravitational potential energy. As the piece of ice rolls down, the gravitational potential energy is converted into kinetic energy, making it accelerate.

The dazzling sunlight becomes hotter and hotter before the piece of ice arrives at the foot of the mountain, and it soon thaws. Isn't this the end of her life? Fortunately, it's not. The piece of ice just converts into molecular energy by absorbing solar radiation energy. Since molecular energy is stronger, the piece of ice is more active, the original cold, hard, square-toed image is long gone, and she becomes a tender water drop.

She flows down into a river where there are many of her siblings. They are mostly from mountains and have undergone similar experiences. The torrential river brings about happy life. She combines with another water drop and gives birth to one boy and one girl.

The daughter comes to a factory, and mankind supplements her energy with electric energy, turning it into purer hydrogen and oxygen. The hydrogen and oxygen, which have chemical energy, release radiation energy in burning and restore their original shapes while providing light and heat to mankind, gaining new life, like Phoenix Nirvana.

The sea refuses no river; greatness lies in the capacity. The son comes to the sea. Mankind finds deuterium in his body, and so helps him undergo nuclear fusion. As he releases great amount of energy in nuclear fusion, he becomes the hero of water.

Without the son, the parents transform into H^+ and OH^-, like the Cowherd and the Weaving Maid [Girl Weaver] (a Chinese legend in which a happy couple become stars separated by the Milky Way. They can meet only once in a year when magpies fly together to form a bridge over the Milky Way.). The two ions combine to form one molecule of H_2O when they meet. The current they form by virtue of their directional motion during the course of their meeting provides electromagnetic energy for mankind.

May the whole family reunite one day!

1.1.2 Amount of Energy

Energy is present in flowing rivers, flying wild geese, falling snowflakes, galloping steeds, blooming flame. Energy takes a value, and to measure it, it is necessary to use the concept of amount of energy.

During the time animal power was used, the carriage was an important means of transportation, and it was from this that the measurement unit horsepower originated. The unit that was invented by James Watt* was used to refer to the power of a steam engine related to a horse's pulling force and was defined as "the work done by one horse which can pull 33,000 pounds and goes at the speed of 1 ft†/min." As our demand for energy increases, we seldom use horsepower except for measurement of the power of an internal combustion engine in the automobile industry and refrigerating effect of air conditioners.

Using different measuring units for different forms of energy is necessary. For heat energy, Calorie‡, Joule,§ or BTU¶ are often used. For electric energy, Wh, kWh, MWh, GWh, or TWh are generally used; and they vary in increments of thousand. Currently, MWh or kWh has been adopted to standardize energy measurement internationally because electric energy is widely used and because it is very convenient for conversion to other energy forms.

In macroeconomy, standard oil equivalent or standard coal equivalent is internationally used for measurement; 1 ton of standard coal equivalent is equal to 0.7 standard oil equivalent. 1 ton of standard oil equivalent is equal to 10 million kcal, 41.84 billion J, 39.65 million BTU, or electric energy of 11.6 MWh.

Earth—Sphere of Energy

The earth itself has gravitational energy, rotational energy, geothermal energy, radiation energy, and attractive energy from other stars, and so can be considered a sphere of energy.

The gravitational energy of the earth mainly refers to the energy produced by earth gravity, which can be converted into thermal energy and kinetic energy.

---

* James Watt (1736–1819), a famous British inventor during the industrial revolution period, created the first steam engine with practical value.
† A foot is equal to 0.3048 m.
‡ The energy needed for 1 g of water to be heated 1°C under atmospheric pressure is determined to be 1 cal.
§ 4.18 J is equal to 1 cal.
¶ The energy needed for 1 pound of water to be heated 1°C is 1 BTU; 1 BTU = 1055.06 J.

The earth, one of the eight planets* in the solar system, revolves around the sun and on its own axis. The inertial centrifugal force produced by the rotation of the earth on its own axis gives the earth huge amounts of energy, which is called rotational energy or kinetic energy. It is calculated to be 2.1×10^{29} J (equivalent to hundreds of millions of times the total quantity of electricity generated worldwide at present) if converted into electric energy.

The inside of the earth can be considered a huge thermal warehouse, storing significant amounts of thermal energy. From the surface of the earth to its center, as the depth increases, the temperature continuously rises, reaching as high as 3700°C at 2900 km underground and 4500°C at the earth center. Underground heat energy is mainly produced by the disintegration of radioactive elements inside the earth.

The external energy of the earth is mainly radiation energy from the sun and attractive energy of the sun and the moon. Solar radiation energy is the main energy on the earth's surface and is the main driving force of surface water and atmospheric motion, which makes the earth surface change its original appearance by weathering and erosion. The attraction of the sun and the moon can produce acting force on the earth and can also convert it into energy. In addition, there are tens of thousands of rivers on the earth running ceaselessly, and they are also sources of massive amounts of energy. To get various mineral resources, people mine them, and hundreds of millions of cubic meters of minerals are moved each year, which can change regional isostasy and is followed by the production of certain amount of energy.

As the earth continues to be acted on by the abovementioned forces, energy is continuously produced and accumulated and then released when it reaches a certain value. There are different ways to release energy. The energies of different forms can be converted into each other, for example, gravitational energy can be converted into thermal energy, which in turn can be converted into kinetic energy, etc. The production and release of energy of the earth are very important for us. This energy will have corresponding consequences and effects on

---

* Pluto was classified as adwarf planet in the fifth resolution passed at the 26th IAU general assembly held in Prague, Czech Republic, on August 24, 2006, and named as minor planet No. 134349, excluded from the nine planets in the solar system. So, there are only eight planets in the solar system.

living organisms in various ways, sometimes producing a large damaging force that changes the ecology of the earth, sometimes producing new mineral resources through motion and changes in the crust of the earth, and so on, just to name a few.

1.1.3 Sources of Energy

Since we have learned about energy and amount of energy, the concept of source of energy comes up when the origin of energy is traced. There are three primary sources of energy: (1) energy from the celestial bodies outside the earth, mainly solar energy; (2) energy that the earth itself contains, such as geothermal energy and atomic energy; and (3) energy produced by the mutual action between the earth and other celestial bodies, such as tidal energy. The concept of the source of energy has evolved into producing the raw materials and resources of energy. In fact, the source of energy is ubiquitous. As far as thermal energy is concerned, if absolute zero* is taken as the yard stick, a temperature difference exists between almost all objects in the universe, even though ice also has thermal temperature. However, there is cost associated with the use of any source of energy. This excessive cost is counterintuitive to the reason for developing and using it, for example, to use the thermal energy of ice for heating supply is diseconomy. Another example, mining the coal thousands of meters underground is not allowed for economic reasons, even if it is feasible technically. Although huge amounts of these energy resources exist, they are meaningless for us since we can't use them. Energy sources take on many different forms and can be roughly divided into seven kinds based on different classification standards (see Figure 1.7).

Classified by whether it could be converted or not: primary energy source and secondary energy source. The energy source that exists in the natural world and can be obtained directly without changing its basic form is called primary energy source and includes coal, petroleum, natural gas, water energy, biomass energy, thermal energy, wind energy, and solar energy. The energy source into which the primary energy source is processed or converted is called secondary energy source and includes electricity, steam, coke, coal gas, and petroleum products. The waste heat and complementary energy are discharged during the course of production in the form of high temperature flue gas, combustible waste gas, waste steam and pressured fluid, etc. No matter how many times the primary energy source is converted, the energy source obtained from it is still called the secondary energy source.

---

* Absolute zero is the lowest temperature in thermodynamics. Atoms and molecules have the least energy allowed by quantum theory at absolute zero. The absolute zero is the zero point of Kelvin temperature scale (denoted by K). 0 K is equal to −273.15°C.

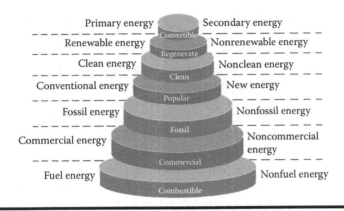

Primary energy / Secondary energy
Convertible

Renewable energy / Nonrenewable energy
Regenerate

Clean energy / Nonclean energy
Clean

Conventional energy / New energy
Popular

Fossil energy / Nonfossil energy
Fossil

Commercial energy / Noncommercial energy
Commercial

Fuel energy / Nonfuel energy
Combustible

Figure 1.7   Energy classification diagram.

Classified by whether it is renewable or not: renewable energy source and non-renewable energy source. The primary energy source can be classified further. Primary energy source that can be renewed is called as renewable energy, otherwise it is called nonrenewable energy. Some examples of renewable sources of energy include wind energy, water energy, ocean energy, tidal energy, solar energy, biomass energy, etc., while coal, petroleum, natural gas, etc., are nonrenewable sources of energy. Geothermal energy, which was originally a nonrenewable source of energy, has now been reclassified as renewable because the earth contains huge amount of this kind of energy on the inside.

Classified by whether it is clean or not: clean energy source and polluting energy source. The energy source that produces little or no pollution during use is called clean energy, otherwise it is called polluting energy. Clean energy generally includes solar energy and wind energy, while polluting energy includes coal, petroleum, etc.

Classified by whether it is popularized or not: conventional energy source and new energy source. The energy source that has been massively produced and is widely used is called conventional energy or traditional energy; it includes renewable water energy resources and nonrenewable resources like coal, petroleum, and natural gas. New energy source is the one that has not been produced massively compared with the conventional energy, including solar energy, wind energy, geothermal energy, ocean energy, biomass energy, and nuclear energy. Conventional energy and new energy are relatively different. Today's conventional energy was a new energy source in the past. Today's new energy source may become conventional energy sources in the future. Coal was a new energy source in the early 19th century when the steam engine was just introduced; petroleum was a new energy source at the end of the 19th century when the internal combustion engine was invented; nuclear energy was also a new energy source in the 1950s when it was just used, but is seen as conventional energy source by developed countries as science and technology develops and many countries are building nuclear power stations, but still it is regarded as new energy source in developing countries.

Classified by whether it is fossil energy or not: fossil energy source and non-fossil energy source. Fossil energy is the energy obtained from fossils of ancient creatures and includes coal, petroleum, and natural gas. Scientists infer that they are the remains of ocean microorganism or animals and plants that were deposited at the bottom of the sea floor tens of millions of years ago, which were later buried underground because of changes in the earth's crust, and then became hydrocarbons* with a complicated structure through decomposition by bacteria and chemical change under high temperature and high pressure action for a long time. The energy sources excluding fossil energy source is nonfossil energy source and includes water energy, solar energy, biomass energy, wind energy, nuclear energy, ocean energy, and geothermal energy.

Classified by whether it is commodity or not: commercial energy source and noncommercial energy source. An energy source that enters the market and is sold as a commodity is a commercial energy source (such as coal, petroleum, natural gas, electricity, etc.), otherwise it is a noncommercial energy source (such as firewood, straw, etc.)

Classified by whether it is combustible or not: fuel energy source and nonfuel energy source. The energy source that is used as fuel, mainly providing heat energy, is called fuel energy source and includes peat and wood, otherwise it is called nonfuel energy source (wind energy, water energy, geothermal energy, and ocean energy).

Tracing the Source—Family Tree of Energy

The Encyclopedia of Energy Source explains that "energy source is the resources that carry energy and directly or through conversion provide it in any of the forms of light, heat and driving force to mankind." At present, there are innumerable data about "energy," its different forms, and its classifications. To understand its origin, we can look into the family tree of energy.

Let's start from electricity. Electric lamps, telephones, computers, fridges, electric fans, electric vehicles, all use electricity. Where does electricity come from? We find power plants along transmission lines. Hydraulic power plants, thermal power plants, wind power plants, geothermal power plants are closely related to these.

---

* The organic compound that consists of carbon and oxygen is called hydrocarbon.

The energy source of hydropower comes from the water with gravitational potential energy, which flows down from a great height and passes through water turbines that convert water energy into mechanical energy, which is then converted into electric energy by the spindles of the water turbines driving the rotors of the generators. Why does the upstream water flow down ceaselessly? The water on the earth becomes steam when sunlight shines on it; which upon meeting the cold air at great heights gathers into clouds when they are supported by ascending airflow; this then becomes rain or snow, which acts as a water source for rivers. The energy source of thermal power originates from fossil fuels, such as coal, whose source is the solar energy collected by plants through photosynthesis. The energy source of wind power stems from wind, whose source is also from atmospheric convection current caused by solar radiation heat. It can be seen that the ancestors of thermal power, hydropower, and wind power are all "Grandpa Sun."

Geothermal power generation is a new power generation technology that uses underground hot water and steam as its driving force. Its fundamental principle is similar to that of thermal power generation. Thermal energy is first converted into mechanical energy, which is then converted into electric energy. The geothermal energy is the natural heat energy extracted from the crust of the earth and comes from the lava inside the earth and exists in the form of thermal energy. Hence, the ancestor of geothermal power generation is "Grandma Earth."

Tidal power generation uses the potential energy of seawater produced by ebb and flow, which occurs as a result of the tide-generating force of the moon causing periodic rise and fall of the seawater and tidal water flow, thereby producing energy. Therefore, the ancestor of tidal power generation is "Goddess Moon."

The energy source of nuclear power generation is nuclear fuel, which is used to produce nuclear energy through nuclear fission or fusion in nuclear reactors. Uranium-235, Uranium-238, and Plutonium-239 are nuclear fuel that can undergo nuclear fission. Uranium-235 exists in the natural world; 1 kg of Uranium-235, upon complete fission, can produce energy equivalent to that of 2400 tons of coal. Regarding nuclear fuel, we only know its present, but don't know its past. Maybe, their ancestor is beside us. See Figure 1.8 for the source and application of electric energy.

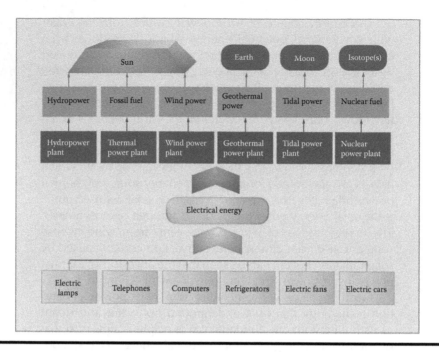

Figure 1.8 Family tree of electric energy.

There are many members in the energy family whose sources are not discussed in detail. If you want to know more about them, please look for the answers in accordance with the above mentioned methods and follow the vine to get the melon.

Chapter 2

Face-Changing Skill: Energy Conversion and Its Law

2.1 Way of Energy Conversion

In daily life, it is not necessary to distinguish the concept of energy and amount of energy strictly; they are almost always regarded as the same concept with regard to nomenclature. The word "energy" is used in the following sections to represent these two concepts.

Energy must be effectively converted from one form to another for use (see Figure 2.1). Conversion of energy in space is transmission of energy; conversion of energy in time is storing of energy. In addition, conversion of energy in form is "face-changing" of energy. The study of using energy is the study of energy conversion and transmission. Although energy forms differ in various ways, every kind of energy can be measured using different methods. Thus, we can determine how much energy is converted from one form to another.

Consider the scenario of a motor running in a car. In this case, chemical energy is converted into thermal energy through the combustion of gasoline or petroleum in the car's engine (internal combustion engine). The thermal energy is then converted into mechanical energy through cylinders that drive the motor, making it run. Therefore, the source of energy of the running motor is the petroleum/gasoline in the car's engine. The driving force of motion in a bicycle is the mechanical energy produced by human physical power pushing the pedals; the source of

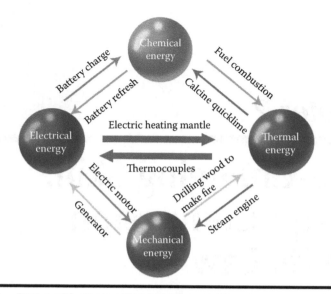

Figure 2.1 Conversion of energy from one form to another.

human physical power is the energy produced upon digestion of the food we eat. Food is the source of energy for all animals. Therefore, without a primary source of energy, there is no energy and driving force.

Human physical power is insignificant compared to the power obtained by energy conversion. For example, 1 ton of coal is needed to drive a train for about 100 km in one hour, which is equivalent to the quantity of work done by ten thousand people in 2–3 days. To perform the same amount of work, using manpower as the source of energy takes longer. Likewise, the journey of a man-made satellite in space is not dependent on conversion of manpower energy conversion. Special equipment or systems are needed so as to convert one kind of energy form into another. The key lies in the advances in engineering technology.

Electric energy can conveniently and effectively be converted into other forms of energy and is the most convenient and commonly used secondary energy. Generally speaking, the conversion of primary energy, such as coal, petroleum, and natural gas, into electric energy occurs in three steps: conversion of fuel to thermal energy, conversion of thermal energy to mechanical energy, and conversion of mechanical energy to electric energy. In the course of conversion, energy can leak out of the edge of a system; this is especially true of heat energy produced, which leaks out from engines, wires, and hot water tanks through radiation and conduction. Some energy will inevitably be lost in conversion; so conversion efficiency is a very important aspect to be considered.

The most common types of energy conversion are as follows.

Conversion of electric energy. The conversion of electric energy into thermal energy is generally through thermal resistance. For example, household electrothermal furnaces produce large amounts of thermal energy through current. Electric energy can be easily converted into mechanical energy in electric machines, and the reverse is also possible, such as in electromotors. In rechargeable batteries, charging restores the chemical energy of the active substance in the battery by supplementing electric energy to the battery.

Conversion of mechanical energy. Mechanical energy can be converted into electric energy through passing through the magnetic field of solenoids, like in electric machinery.* Mechanical energy is converted into thermal energy when an object produces heat through friction, for example, when a match is ignited by the heat produced by the friction of quickly striking it against the matchbox.

Conversion of thermal energy. Thermal energy can be converted into electric energy by using temperature difference to produce electric potential. For example, the thermoelectric couple of a thermometer uses two conductors made of different materials to form a closed loop circuit that produces electric energy when the temperatures at the two ends of the junction are different. Thermal energy is converted into mechanical energy. For example, water that is heated produces steam, which drives machinery, thus producing mechanical energy. Humans have used this method to convert thermal energy into mechanical energy since the steam engine was invented. Thermal energy can be converted into chemical energy when limestone is burned in a limekiln at over 900°C to become quicklime, whose energy will be released when it meets water.

Conversion of chemical energy. Chemical energy is converted into electric energy when positively and negatively charged ions, respectively, converge at the anode and cathode of a battery—this is what happens when a battery is charged. Chemical energy is converted into thermal energy when fuel produces heat in the course of combustion.

Mayer—How Is Energy Converted?

The law of energy conversion and conservation was one of the three major discoveries of the 19th century in the field of natural science. The man who first discovered conversion of energy was Julius Robert Mayer,† a doctor, who was often called a "madman" (see Figure 2.2). In 1840, he began to

---

* Electric machinery is a kind of electromagnetic device that converts or transmits electric energy based on the law of electromagnetic induction.

† Julius Robert Mayer (1814–1878) was a doctor and physicist, from Hamburg, Germany. He found and stated the law of energy conversion and conservation.

Fr. Berrer.   Heilbronn.

Figure 2.2 **"Madman" Julius Robert Mayer.**

practice medicine on his own in Hamburg, Germany. Mayer was always curious and carried out observation, research, and experiments on anything that caught his interest.

In 1840, he came to India along with a fleet of ships in his capacity as a doctor. When the fleet landed at Calcutta, the seamen all took ill because they were not acclimatized. Mayer treated them by bloodletting, a therapy in accordance with the old method of bleeding bad blood out. In Germany, the doctor just needed to run a needle into the veins of a patient to cure the disease—darker red blood would run out from the veins.

But at Calcutta, when the same procedure was done, red blood ran out from the veins. Mayer then began to theorize that human blood was red because it contained oxygen, which helps human bodies produce heat to maintain temperature. The weather at Calcutta was hot, and so not much oxygen was needed to maintain the temperature of the human body; therefore, the blood in the veins was a brighter red than that observed in Germany.

Where does the heat in the human body come from? A human heart weighs 500 g at most; its pumping action cannot produce the heat necessary to maintain the temperature of the human body. It seemed that the temperature of the human body must be maintained by the blood and flesh of the body, which came from the food eaten—food from plants or meat from animals—which grew by absorbing energy from the sun. Where did the light and heat of the sun originate from? If the sun was a piece of coal, it, of course, could not have been burning for such a long time. The sun must have another source of energy. The light and heat of the sun must therefore come from an energy unknown to us. Mayer boldly speculated that the temperature at the center of the sun was about 27.5 million degrees (now we know it is 15 million degrees); so the sun must have tremendous energy! Mayer thought more and more, deeper and deeper, and finally summarized his musings in one question: how is energy converted?

He wrote the paper, *Remarks on the Forces of Inorganic Nature*, as soon as he returned to Hamburg. He used his own method to measure the thermal equivalent of work[*] and found it to be 365 kg m/kcal. He contributed his paper to *Annalen der Physik* (*Annals of Physics*), but it was refused. He had to publish it in a medical magazine that was little known.

Afterwards, Mayer tried to popularize his theory by delivering speeches: "You see, the sun radiates light and heat, the plants on the earth absorb them and produce chemical substance ..." But, even the physicists of his time did not believe him. They disrespectfully called him a "madman." Later, even his family doubted his sanity and fetched a doctor to cure him. Mayer was not only misunderstood academically, his personal life was a torment too—his youngest son died and his younger brother was implicated in revolutionary activities. Mentally broken from this series of misfortunes, Mayer jumped from the third floor of

---

[*] Thermal equivalent of work refers to the quantitative relationship between the unit of heat (calorie) and the unit of work, equivalent to the quantity of work in unit heat. At present, the international system of units stipulates that Joule be used as the unit for heat and work, and so thermal equivalent of work has lost its meaning.

a building in 1849. However, he did not die; he broke both his legs and became crippled. He was sent to Goettingen Mental Hospital where he was tortured for 8 years.

The world did not discover Mayer and recognize the value of his theory until 1858. It could be said of Mayer, in his later years, that after great suffering comes happiness. After he left the mental hospital, he was conferred an honorary doctorate by the Academy of Natural Sciences, Basel, Switzerland, he received the Copley Medal of the British Royal Society, and he became an honorary doctor of philosophy of the University of Tuebingen, Germany, and an academician of the Turin Academy of Sciences, Italy. Although Mayer did not finalize the formulation of the law of energy conversion and conservation, the great physicist and doctor, who was not accepted because of his theory that surpassed his times, always persisted in his search for truth, for which he won the respect of and is remembered by generations of scientists.

2.2 Law of Energy Conversion and Conservation

The law of energy conversion and conservation is also called the first law of thermodynamics. It states that energy is neither produced out of the void nor does it disappear into the void; it can only be converted from one form to another, or from one object to another, during the course of which its total amount does not change. Whenever the energy of one form or object reduces, the energy of another form or object will increase by the same amount. The sum of all the different kinds of energies in a system will not change as long as there is no energy entering in it, whether the energy in it changes gradually or suddenly. Although the discovery of energy conversion and conservation was inevitable as human knowledge of the law of natural sciences increased to certain extent, the journey to the discovery was still tortuous, hard, and exciting.

From the 19th century onwards, the Industrial Revolution, marked by the widespread use of the steam engine, swept across Europe. But the principle of the conversion of thermal energy into mechanical energy was not known at that time, and neither was the theory of the work behind them. Engineers groped their way through past experience. Carnot[*] raised the theory of heat machinery,[†] later known as A Carnot Heat Engine, in 1824, laying the first theoretical foundation for thermodynamics.

---

[*] Sadi Carnot (1796–1832), a French engineer, one of the founders of thermotics, put forward the concept of "ideal thermal machinery."

[†] A Carnot heat engine is an engine that operates on the reversible Carnot cycle, acts by transferring energy from a warm region to a cool region of space and in the process converting some of that energy to mechanical work. It is the most efficient thermal machinery among all machinery working between the same high temperature heat source and low temperature heat source in theory.

In 1840, Mayer concluded that "the input and output of the energy of all forms in human body are balanced, it is constant in quantity," when he studied the conversion of chemical energy and thermal energy in the human body. In 1842, he expanded the conclusion to energy sources outside the human body and expressed his idea of energy conservation in the physical and chemical world as well.

Joule* described and demonstrated the quantitative relationship of the conversion of electric energy into thermal energy in 1840. He undertook a series of experiments from 1843 to 1847, made measurements on the whole aspect of mechanical energy and conversion between electric energy and thermal energy, and provided the value of the thermal equivalent of work for the first time. Joules's experiments, and ultimately the determination of the thermal equivalent of work, indicates that the energy in the natural world is indestructible and that heat is just a form of energy. Joule published his experimental report in 1843 in *Philosophical Magazine* which showed that the value of the thermal equivalent of work was 423.9 kg m/kcal, that is, the thermal energy per kilocalorie could be converted to mechanical energy of 423.9 kg m; this work was not recognized by the scientific community immediately.

Helmholtz,[†] a German physicist, first expressed the conservation of mechanical energy of a separate system in mathematical form in *Conversion of Force* and then expanded it to thermotics, electromagnetism, astronomy, and physiology, expounding his idea systematically and explicitly stating that all forms of energy can be converted and conserved.

Joule published the law of energy conversion and conservation in 1853. This law was the first great proof of the inherent unity of all natural sciences. It provided a solid theoretical foundation for the technical progress of all kinds of energy power machinery, broke through the illusion of the "perpetual motion machine" popular at that time, and promoted the development of the Industrial Revolution. It revealed the universal relationship between the different forms of the material world and the universal relationship among all branches of natural sciences. It has become the most widely used powerful weapon to discover and transform the world.

Does the Perpetual Motion Machine Never Stop?

The concept of the perpetual motion machine originated from India and spread to Europe in the 12th century. The earliest and most famous design was that by Honnecourt[‡] in the 13th

---

* James Prescott Joule (1818–1889), a British physicist, made great contributions to thermotics, thermodynamics, and electricity.
† Hermann von Helmholtz (1821–1894) was a German physicist, mathematician, physiologist, and psychologist.
‡ Villard de Honnecourt was a 13th-century artist from Picardy in northern France. He is known to history only through a surviving portfolio or "sketchbook" containing about 250 drawings and designs of a wide variety of subjects. (Note From Wikipedia).

century in Europe. Many European people continued to study and invent their own perpetual motion machine.

Leonardo da Vinci[*] (1452–1519) also created a perpetual motion machine (see Figure 2.3) during the Renaissance. He believed that the wheel would ceaselessly rotate along the direction of the arrow as long as the heavy balls on the right side were designed to be farther from the wheel center than the heavy balls on the left side. In fact, it can be determined from the lever balance principle that although the rotary acting force of each heavy object on the wheel was large, the number of heavy objects was less, and so the reverse rotary acting force (torque) of the heavy objects on both sides of the wheel was equal, offsetting each other and balancing the forces on each side. However, finally, the rotating wheel became motionless because of the action of friction.

Buoyant force was one of the main factors used in perpetual motion machinery design in the 19th century. One famous perpetual motion machinery design is that of a line of iron boxes installed on the wall of a tower, which could rotate like a chain. The iron boxes on the right side were put in a container full of water. The designer thought that if there were no containers on the right side, the chain would be balanced because the number of the boxes on both sides was equal. The iron boxes on the right side would go upward under the push of water when

Figure 2.3   Leonardo da Vinci's model of the perpetual motion machinery.

---

[*] Leonardo da Vinci (1452–1519), an Italian Renaissance polymath, painter, inventor, scientist, and architectural and military engineer.

they were immersed in the water, thereby obtaining buoyant force, driving the line of boxes rotating around the wall of the tower (see Figure 2.4).

This machine was also not successful. This was not because of technical difficulties but because of design. When an iron box passed through the bottom of the container, it would bear the pressure from the water above, similar to that experienced by the base of the container, which would offset the buoyance of the iron boxes above, so the perpetual motion machinery could not move perpetually.

Magnetism also played a role in the "perpetual motion machinery myth." The "magnetic perpetual motion machinery" in Figure 2.5 was designed by John Wilkins[*]. He put a powerful magnet A on a small pillar and rested two oblique wooden grooves M and N on the pillar. There was a small hole C on the upper wooden groove M; the lower wooden groove was curved. The inventor thought that if a small iron ball B was put on the upper wooden groove, it would roll upward under the attraction of the magnet A, fall on the lower wooden groove N when it rolled to the small hole C, continue to roll to the lower end of the groove N all the way, and then go upward to the groove M

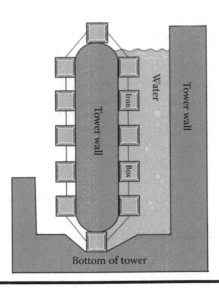

Figure 2.4   The perpetual motion machinery using buoyance.

[*] John Wilkins FRS (14 February 1614–19 November 1672) was an Anglican clergyman, natural philosopher and author, and was one of the founders of the Royal Society. He was Bishop of Chester from 1668 until his death.

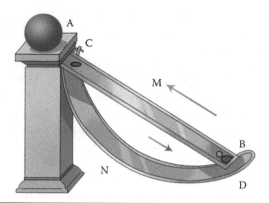

Figure 2.5   Magnetic perpetual motion machinery.

along the bend D, with the entire process being repeated under the attraction of the magnet A. In this way, the small ball would ceaselessly roll and never stop. But, this design also failed.

Some people made designs that used the inertia of the wheel and capillary action of fine tubes to get an effective driving force. But all these designs failed. In fact, in all designs of perpetual motion machinery, a balanced point was always found where all forces were offset and there was no driving force to make the perpetual motion machinery move. So, all the perpetual motion machinery had to stop at a balanced point and become motionless.

"Perpetual motion machinery" attracted the attention of many eminent people even though the basic laws of nature had not yet been fully mastered. No perpetual motion machinery was ever made successfully, nor did its designs pass scientific inspection, though its imagination fueled the intelligence and wisdom of many people. Perpetual motion machinery is just an illusion; it can never be successful because it goes against one of the most fundamental laws of the natural world: the law of energy conversion and conservation.

Chapter 3

Like the Shadow Following the Body: Ubiquitous Energy

3.1 Energy and "I"

Energy is an integral part of everyday life, but people usually ignore its importance. Life lies in movement, and energy is its foundation. Life needs energy every day; just as a steam engine needs coal, an internal combustion engine needs gasoline, or an electromotor needs electricity, energy is the fuel of your body. Even if you do nothing, functions like breathing, heartbeat, blood flow, body temperature, and so on require energy. As long as you are alive, you need energy, for which you have to eat. Every mouthful of food you eat contributes to the continuity of your life. The more physical activities (work and exercise) you indulge in, the more energy you consume. When the energy you take in is more than what you consume, it will be stored in your body and become fat, which is like depositing your money in a bank. When the energy you require is less than what you consume, it will be taken out from the fat in your body for consumption, which is like spending your previous deposits.

In addition to the smooth working of your body, energy is indispensable in your life. When you are sitting on the sofa and turning on your television, a thermal power generator is converting the thermal energy of coal into electric energy that transmits TV images and voice. When you call one of your friends through your mobile phone, the chemical energy stored in it is converted into electric energy. When you attend a party by car, petroleum is converted into mechanical energy through the car's motor. When you and your relatives and friends sing with microphones, when you see a film in a cinema, when you choose a new dress in a supermarket, when you go upstairs standing in a lift, when you turn on a table lamp for reading…, energy accompanies you imperceptibly, like the shadow following the body. Energy is so important that no one can do without it, as it is necessary for basic processes like food, clothing, shelter, or transportation.

When you work, energy is even more important. The first thing a lot of people do when they arrive at their offices is to turn on their computers to get information, receive and send emails, dispose documents. Without energy, the computer is just a decorative piece. You get in touch with your colleagues and clients using telephones, print documents with a printer, duplicate documents with a copier, fax documents with a fax machine, go on a business trip by plane, train, or ship—you cannot do all these things without energy.

Without a doubt, humans are more comfortable now compared with life 100,000 years ago. Over the years, means of transportation have changed from carriages to cars, trains, and planes; lighting sources have changed from campfire, candles, and kerosene lamps to incandescent lamps, neon lamps, and energy-saving lamps; communication and interaction methods have changed from letters to video chat. Many people long for an idyllic romantic life without the modern trappings of civilization. They leave the modern world to live a life in tune with nature (with birds and beasts, insects, and fish); they begin work at sunrise, stop work at sunset, grow crops in spring, and harvest them in autumn using slash-and-burn farming methods. But most of them are unable to survive like this for several days and so return to modern society. It is the very progress of civilization and development of energy use that provide everyone with a faster, more convenient, and colorful life that is rich filled with future possibilities.

The Days without Energy

Let us imagine how it would be if energy disappears.

Energy disappears when I wake up. The electronic alarm clock, which is most annoying every day, is silent at last. I open my eyes, sluggishly take up my handset whose screen is black. I do not know what time it is. Probably, it is getting late. I get up, turn on the faucet, no water runs out. I cannot wash my hands or face, and even worse, I do not dare go to the toilet. I go out the door, go downstairs, leave my lovable car in the garage, ride my dirty bicycle, and by standing in a long queue visit the public toilets on the way to my company.

I finally arrive at my company and do not know how late I am—anyhow, the fingerprint attendance machine broke earlier. Looking up at the towering office building, I despair— my office is on the 33rd floor. I climb up with my colleagues step by step, each one encouraging the other; none can make a joke of others' being late for work. I sit down, open my computer unconsciously—without any reaction. Looking at the display, which is just like a mirror, I hear my colleague in the administrative department informing every one of every floor to attend a meeting. And then over 100 people gather at the meeting room. The boss, standing at the rostrum, loudly calls everyone to persist in doing good work. Without a microphone, his voice is drowned by the employees' whispering. I return to office after the meeting, look at the phone which has not rung even once in one hour, and suddenly have the illusion that I am the only one left in the world. I only vent my loss by fretfully pacing the room as sweat runs down my face.

Everybody's lunch is bread and water, which was stored by the company earlier for emergencies. In the afternoon, the office is even stuffier; the thermometer reads 36°C. The thunderstorm of yesterday evening brings little coolness to the city. There is no wind; some clouds float in the blue sky, but they cannot shield us from the scorching sun. I take up anything at hand to wave and create some cool breeze. I suddenly remember my appointment with a client to confirm contract details. I reach toward the phone by habit, but my hand stops in mid-air. I have to go down the 33 floors of my office building with several copies of revised manuscripts of the contract and ride my bicycle in the burning sun. The 2-way 10-lane highway becomes the arena of various bicycles and even roller skates. Along the way, I see there are many people in need of money in front of closed banks; shopping centers close business, and peddlers sell candles at a high price at roadsides. I stop my bicycle and buy two packs…

I ride my bicycle home. As the sun sets, the world is shrouded in blackness. After I eat bread, I have nothing to do because there is no energy. I face the silent computer and TV, I have no way of spending time. I blow out the candle and lie down on my bed but cannot sleep. I look at the dark city outside the window, as if I am awaiting the end of the world. As time passes by, I only pray for the coming of tomorrow, minute by minute.

3.2 Energy and "Us"

Energy is the physical basis for human activity, and social development is not possible without high quality of energy forms and advanced energy technologies. We have been using energy even before civilization started. With continuous scientific development and the springing up of new technologies, energy will be more widely used and will play a greater role in our daily life.

Petroleum is the major energy source, a prime constituent of the overall primary energy consumption in the world, and its products include liquefied petroleum gas, petrochemical materials, fuels, and lubricating oil stock to pitch. The processing of petroleum also releases petroleum gas in great quantity. Highly efficient, economic, and properly used liquid fuels that are mainly divided into internal combustion engine fuel, boiler fuel, and kerosene can be obtained after the petroleum is processed. Other petroleum products including lubricating oil, wax, pitch, and petrochemical products, such as petroleum solvent, ethylene, propylene, polyethylene, are also obtained.

Natural gas is a kind of mixed gas; its main component is methane, which is characterized by its easy combustibility, cleanness, and high thermal value; it also leaves no ash upon burning and it is pollution free. Like petroleum, natural gas is also a very important organic chemical material. It requires less investment, is highly efficient, and can adapt to sudden charge change and can be used to heat boilers to produce steam. Productivity can be increased by 30% if natural gas is used instead of coke. Many substances separated from natural gas are very basic chemical materials from which lots of chemical products, such as synthetic fiber, synthetic rubber, synthetic plastic, and chemical fertilizer can be made. These chemical products have many advantages such as widespread application, low cost, high output, and fast development. Hence, their use is very important for economic development and social life.

Since the end of the 18th century, when the Industrial Revolution began, coal has been widely used as the fuel in the industry; it was an important source of energy apart from petroleum and natural gas. With the invention and use of the steam engine, coal brought an unprecedented huge productive force to the society, giving an impetus to coal, iron, chemical industry, mining, metallurgy and other industries. Owing to its high heat, rich reserves in the earth, and wide distribution, coal is generally easy to mine. In addition to being used as industrial fuel to obtain heat and kinetic energy, what is more important is that coke (for metallurgy) and coal tar can be made from coal. Besides, various chemical products can be made from coal through chemical processing. Therefore, coal is a very important chemical material. In China, many medium- and small-sized nitrogen fertilizer plants produce chemical fertilizer using coal as a raw material. Coal also contains lots of radioactive and rare elements, such as uranium, germanium, and gallium. They are important raw materials for semiconductor and atomic energy industries. Hence, coal plays an important role in modern industries, no matter if it is heavy industry, light industry, or energy industry, metallic industry, chemical industry, mechanical industry, or textile industry, food industry, transportation industry, etc. All industrial

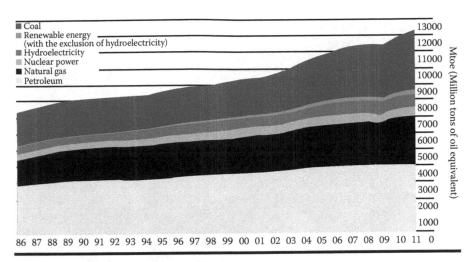

Figure 3.1   Global energy consumption.

sectors consume a certain amount of coal to some extent; so coal is called the "real food" of industries.

In 2011, the human population rose past the 7 billion mark—an increase of 1 billion compared to 1999. Simultaneously, energy consumption increased by 16 times. In recent decades, energy supply is unable to always meet the energy demand, although energy development and production scale are constantly being increased.

At present, fossil fuel is the main source of energy. China and a few other countries use coal as the main fuel, while other countries use petroleum and natural gas. Based on present consumption, petroleum and natural gas can only last half a century at most, while coal can only last for 200 years at most. Global energy consumption data given by the British Petroleum Company is given in Figure 3.1. To maintain sustainable development of economy, ecological balance, and progress of civilization, we must thoroughly change our considerable dependence on fossil energy.

Indian Grid Disturbance Affected 600 Million People

Energy and social development are closely related to each other in the present world. Daily routine will be seriously affected, and even stop, if energy is interrupted (see Figure 3.2).

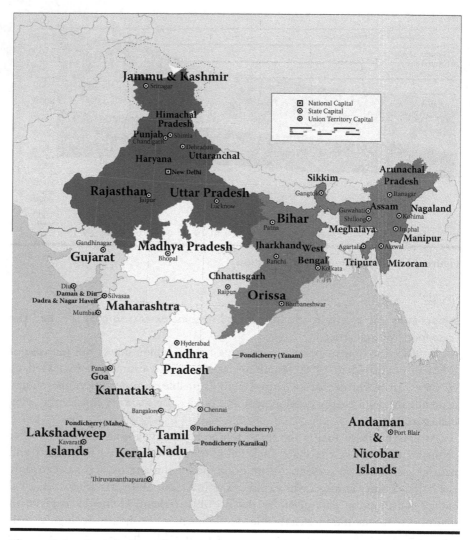

Figure 3.2 On July 30 and 31, 2012, two severe power blackouts affected most of northern and eastern India. Indian states: dark gray area: affected 2 days by the power outages (on 30th and 31st July); light gray area: affected 1 day by the power outages (on 31st July). Date: 02:48, 30 July 2012 (+05:30) to 20:30, 31 July 2012 (+05:30). Location: Northern India.

Beginning from one o'clock in the afternoon on July 30, 2012, three Indian grids were paralyzed one after another. Nearly half of India, including the northern and eastern regions, most of the northeastern region, and New Delhi, covering 22 provinces and cities, fell into darkness. The official postanalysis statistics showed that the grid disturbance had affected

680 million Indian people, caused an economic loss amounting to hundreds of million US$, and set the record in worldwide blackout affecting a population. Some commented: "One power failure, even for a few seconds, brings about damage that can be compared with an earthquake."

There was chaos on the roads, and trains stopped operating. The power failure first hit the northern and eastern regions of India. The New Delhi subway system completely stopped operation after power failure. The trains stopped in dark tunnels; lots of passengers walked out from sultry subway stations at lunch time and jumped into crowded buses. The traffic on highways was a mess because the traffic lights had stopped working; serious traffic jams occurred during morning peak hours; and traffic policemen directed vehicles by themselves. Besides, railway transportation was heavily hit. The spokesman of the Indian Railways said that about 500 trains were stopped or were late because of the blackout (traffic light stoppage), and so a great number of passengers were detained and railway traffic was in a mess …

The blackout also led to a partial pause in Indian public service. A crematorium in New Delhi had to cremate three remains by burning firewood after the blackout. A clerk of a branch of Bank of India said that the blackout paralyzed the main server of the bank and that service had to be paused. Amitabha, an intern of Hong Kong and Shanghai Banking Corporation, said: "It's a shame, power is the very basic of facilities; this situation should not have occurred at all." In Calcutta, a big city in eastern India, some governmental departments were forced to stop work earlier; some office buildings used kerosene generators to generate electricity; hospitals and airports were forced to use reserve power supply.

Elevators stopped, and so miners were caught. In Burdwan, 180 km in the northwest of Calcutta, more than 200 miners of the Eastern Coal Field Company were stranded underground because the elevators did not work after power failure. The general manager of the company said that they could not lift them to safety unless power supply was resumed. The miners were requested to move to places with better ventilation. Rescue workers provided food and drinking water to them. The Chief Minister of West Bengal requested governmental agency staff to go home so that the reserve power supply could be sent to the mines to help in rescuing miners.

The public suffered from power failure. Their normal life was seriously affected by the blackout. "I arrived at Jaipur by train after 2 o'clock in the morning; it was dark when I went out from the station. I rode a rickshaw to the hotel. There was no road light all the way; many people were sleeping at the roadside; it was so hot," said the one local resident.

The temperature on that day in New Delhi was about 35°C, humidity was 81%. Lots of residents awoke because it was too hot, finding their fans and air conditioners had stopped and that the whole city was dark. Manjali Mishra grumbled: "This is a very bad night. It is damp; the electric fan can't be used, it is really a tough time. The day is similarly disastrous, no electricity, no water."

The Indian student Sliba Narang said, "It is the abhorrent blackout that ruins everything." On Monday afternoon, Saliba Narang, an Indian senior high school student aged 17, was waiting for the school bus to go home after school, but it did not appear for a long time. Wanting to go home quickly, he rushed to the subway station, but the trains had stopped running. He sighed and returned to the classroom to wait. He did not go home by train until evening when the power supply in New Delhi was resumed.

Jayao, 54, was on the way to the taxation department by subway to submit his income tax returns. The train stopped in the tunnel without warning, and he missed the deadline of summiting the form. "Today is the last day, what shall I do?" He was upset, but had no choice except to wait for rescue.

Originating from Fire, Our Footprints

Human civilization began with the burning and lighting of kindle. From fear to familiarity and usage of fire, human beings entered the era of consciously utilizing and improving energy. We have developed beyond the hunting and gathering civilization, and the domestication of the Agrarian Age. Coal and steam from the Industrial Revolution, together with electromagnetic energy from modern society, are continuously providing the driving force for the evolution of human civilization.

Chapter 4

Burning Passion: Beginning of Greatness

4.1 Discovery and Utilization of Fire

Fire is a phenomenon in which light and heat is radiated during combustion. It is also a pattern of energy release. Combustion is a drastic chemical reaction releasing light and heat, which can only occur in the presence of a comburent, combustible material and when the burning point is reached at the same time. Plants on the earth absorb solar energy and carbon dioxide while releasing oxygen—the comburent—through photosynthesis, and providing large amount of wood and straw—the combustible material. However, combustion cannot happen even if both these substances exist, because a burning point has to be reached to trigger the drastic chemical reaction. Nature is so amazing that lightning in the sky and the eruption of volcanoes provide burning points for the peaceful earth and ignite wood and weeds, which adorn the earth with splendid color.

The oxygen produced in photosynthesis is active and has many chemical properties. It can not only drastically react with combustible materials but also slowly react with most elements (rare gases and hardly active metal elements such as gold, platinum, and silver are the exception). This process is called oxidation, and the reaction rate is dependent on elements. Digestion system in the animals' body and rusting of ironware are relatively slow oxidation processes.

Fire is often called the man-made sun. Its strong light illuminates dark roads and caves. Burning flame protected us from the fierce beasts that roamed about. Warm campfire helped us get through cold winters. Forest fire gave us land for farming and habitation. Fire not only changed the external environment, but also made human beings exclusive members of delicatessen restaurants with cooked food, which promoted the human physiological and genetic evolution to a rate much faster than that of other animals.

Discovering, and mastering the use of fire was a great breakthrough. Previously, humans used clubs and stone axes to assist them in their ventures (see Figure 4.1). These tools only depended on physical torque, relying on fundamental energy source from human bodies. Later, human domestication led to the release of large amounts of energy, which originated from solar energy and was absorbed by organic matter directly or indirectly through photosynthesis. The human hunting and gathering civilization lasted from 3 million years ago to 12,000 years ago, longer than any of the later civilizations. Firewood-based fire was the most important energy source at that time.

Everything was difficult in the beginning. Our ancestors' domestication of fire was of course not as simple as striking a match. The most precious burning points—volcanic eruptions and lightning—were very dangerous, and so humans were wary of them and approached them cautiously. The price our ancestors paid to tame and own fire was bigger than that of any sacrifice of later-day humans in their exploration of the world around them. History cannot be repeated, but we can imagine their hardship and admire and appreciate their ingenuity.

At times, volcanic eruption destroyed our ancestors' territory. Some people were even buried by scalding magma before they could escape; the survivors ran everywhere and did not know if the lighted wood would become magma and burn everything on sight. Gradually, people seeing that magma ignited wood, began to

Figure 4.1 Hunting by hominids.

Figure 4.2 Some tribes still retain the habit of drilling wood to make fire.

try picking up the magma-ignited wood but dropped it instantly after they were scalded. However, some happened to grab the unignited part and took the first step to domesticating fire. Over countless generations, people explored and pondered the possibility of keeping the fire going, without it burning out. Lightning often occurred, but it was fleeting and did not give enough time to humankind to light their kindle. We do not know whether humans learned to use fire from these volcanic eruptions or the frequent lightning lighting up the sky of prehistoric earth.

After countless incidents of trial and error, humans finally stopped relying on natural fire (volcano and lightning) and discovered the method of generating fire by rubbing combustibles together quickly or striking stones against each other or scraping them with big knives (see Figure 4.2). This is a joint use of physical and chemical methods and, perhaps, is a greater discovery than any of our modern complicated works.

The Legend of Making Fire by Drilling Wood

Fire has existed in nature since a long time ago in the form of volcanic eruption and forest fires caused by lightning. At the beginning, our ancestors did not know how to use fire, so they ate raw fruits from plants and uncooked meat from animals. They were quite afraid of fire when they first saw it. Later, they occasionally picked up beasts burned by fire, ate them, and found them to be delicious. They gradually learned to cook food on the fire to eat and managed to maintain fire.

Nevertheless, maintaining fire was time-consuming and strenuous; the fire would often go out, so making fire was the most difficult and urgent problem to solve. It was then that the ancient legend of drilling wood to get fire first appeared. The principle behind it is the generation of heat by friction. One end of a small wooden pole is put on a large piece of wood and quickly rotated to generate sparks at the point of interaction. Since friction will convert kinetic energy to heat energy, hay, fur, or other easy-combustible materials can be ignited, thus creating artificial fire.

Sui-Ren Shi* from China, more than 6000 years ago, drilled wood to get fire, taught people to eat cooked food, and ended the history of Chinese ancestors' eating the raw flesh of and drinking the blood of animals. He once saw a woodpecker looking for a worm in the hole of a tree with its long and pointed beak. The worm was hidden deeply in the wood, so the woodpecker had to peck the wood for a long time, and accidentally generated thick smoke and fire. Sui-Ren Shi, it is said, was inspired to make fire using a similar method and started the epoch of drilling wood to get fire. This was an extraordinary invention. From then on, people could eat cooked food any time. Sui-Ren Shi also taught people to catch and cook fish, soft-shelled turtles, mussels, clams, and so on with fire. These were previously inedible as they had a fishy smell when eaten uncooked—fire added to the range of food that could be eaten by humans. In the course of making fire, people found that not all wood could be drilled to generate fire. It was also necessary to choose different wood according to the seasons. If a piece of wood was randomly selected for drilling, it might not generate fire. In spring, dry elm or willow are necessary; in summer, jujube wood, apricot wood, or mulberry wood should be used; in autumn, people need to find oak or wine goblet; in winter, wood of the Chinese scholar-tree or sandalwood was used. In the period of the Yellow Emperor Xuanyuan, there were special officials set up in each part of China to manage fire—they were in charge of choosing the wood that could be drilled to make fire.

---

* Sui-Ren Shi, named Yunruo, is the inventor of fire in ancient China; he was the first of the three ancient emperors. *Pick up the Lost* records: "there was a large tree in Suiming state, named Sui, covering a large area. Later, a saint traveled to the state, a bird pecked the tree, fire suddenly came out. The saint was inspired, so he drilled wood to make fire, was called as Sui-Ren Shi."

4.2 Cooking and Human Evolution

One critical function of fire is improving human living environment, as it is used as a source of light, heat, resistance to fierce beasts, etc. It also of course helps in cooking.

Compared with chimpanzee and ape bodies, humans have larger cranial capacity; smaller teeth, bone structure, and muscular tissues; and shorter intestines, which is closely linked with the process of cooking. The cooked food is easier to chew and absorb; the organic chemical emissions gives out the inviting smell of fat and other hydrocarbon; bacteria and viruses are annihilated by high temperature, which helps in prevention and reduction of diseases. No other animals have learned cooking so far. This unique skill has accelerated human physiological and genetic evolution to speeds faster than other animals, and this is represented by smaller teeth, shorter intestines, and bigger cranial capacity.

Cooking assists the digestive process occurring in the mouth, stomach, and intestine. Hard shells, firm skins, tendons, and muscles, when heated to high temperature during cooking, change their physical and chemical properties and become easy to chew and bite. Taking in raw food depends on the digestion ability, that is, on gastric acid and peristalsis of the stomach. However, with the powerful support of the 1000°C fire, the digestion becomes more comprehensive and efficient; the intestine absorbs higher quality nutrition, saving our physical energy. Nowadays, we literally cannot eat raw rice or meat if they are not cooked. Thus, it can be seen that cooking is equal to the auxiliary process of chewing, digesting, and absorbing food; it makes us optimize and simplify parts of our bodies—teeth, stomach, and intestine.

What is the relationship between larger cranial capacity and cooking? People regard food as their basic necessity. Before they learned to cook, humankind spent most of their waking time eating. Cooking helps us eat faster, better, and more easily; consequently, we have more time to do other things. Chimpanzees spend 6 h every day chewing food; humans spend only 1 h on this since they have learned to cook. Humans obtained a valuable time advantage within 4 million years—5 h/day—in the natural competition of survival. What did they do with the extra time? They started thinking: what is the relationship between mountains and rivers? What are the laws governing day and night and the change of the four seasons? How could our shelter be built to be more comfortable? How do we distribute the labor force of gathering fruit, hunting, and cooking more effectively? More mental work assists the human brain to be fully used. Therefore, the human cranial capacity became bigger and bigger so as to meet the demand of heavier mental work. Our cranial capacity today is three times larger than it was 4 million years ago.

In this sense, cooking not only makes food more delicious but also switches the form of our using natural energy to a new platform and evolves our brain, stomach, intestine, teeth, and other organs more quickly. Human beings became neither herbivores nor simple carnivores but "food-cooking animals." The increase of the number of human sites of the late Old Stone Age convincingly shows that humankind lived much easier because of the widespread use of cooking. Besides, population, exchange of ideas, inventions, and explorations also increased. With the development of cooking, the human social attribute continuously intensified; people needed closer cooperation and subdivision of labor; dietary time became more regular; dwelling places became more fixed; communication became fuller; and human relations became firmer.

Human civilization started with fire, and over time this has grown to encompass various forms, such as from the candles on birthday cakes to campfires held in the open countryside, from fireworks of festival celebrations to the Olympic flame, etc. All these transitions show human's respect and love for fire—the great energy form.

The Chimpanzee Also Performs the Flame Dance

Anthropologist Jill Plutz* said that animals' understanding of fire includes three steps: The concept of fire, making fire, and controlling fire. Most animals fail in the first step because of instinct. Reed frogs in West Africa will stampede once they hear the sound of fire crackling, Australian bettongs will panic when they hear the sound of fire, but Plutz finds that the wild chimpanzees in the savannahs of Senegal will calmly perform a flame dance in front of the fire.

Wildfire often occurs on the grasslands in the savannah region when the dry season is about to end. One day in 2006, Plutz stayed with the chimpanzees that were being researched even though there was threat of wildfire. To her surprise, the chimpanzees were not afraid of the fire and calmly followed behind her to pass it; and their male leader even ritually performed a flame dance in front of the fire. They even accurately predicted the direction of the raging fire. Once, the flame surrounded them on three sides; yet, the chimpanzees kept calm and figured out how to escape from the flame circle safely, and this astonished Plutz.

Some people believe humans were born with the fear of fire. To eliminate this fear of fire is the first step of learning to make fire and finally control it. The wild chimpanzees' flame dance before the fire shows that they are close relatives of

---

* Jill Plutz was an anthropologist from Iowa State University, IA.

humans. Their action is enough to distinguish them from non-human animals and provides valuable clues and a powerful incentive for researching when our ancestors took the key step—learning how to control fire—in the course of human evolution.[*]

---

[*] Refer to *The Independent* dated January 16, 2010.

Chapter 5

Domestication a Godsend: Exploring Farming Force

5.1 Origin of Domestication

The human agricultural civilization lasted from about 12,000 years ago to AC 1500. However, the origin of domestication was much earlier and lasted much longer. Domestication refers to the process of making wild animals or plants fit for domestic life or cultivation. We usually construe domestication as pertaining only to animals, such as domesticating wolves to dogs, wild boars to hogs, wild horses and camels to mounts, etc. In fact, crops can be domesticated too. For example, wild wheat was collected for seeds and sowed in our land, watered, fertilized, and then sowed again at relatively regular intervals of time in relatively concentrated areas to produce more grain. In this way, wild plants were domesticated into crops. By the late Old Stone Age, humankind had a large population who fed on wild animals and plants and became hunters and gatherers. Thousands of years later, most of the increasing population cultivated plants and raised animals—the hunters and gatherers of the olden era had now become farmers. The history of crop cultivation is different in various regions of the world; it began in around 6500 BC–3500 BC in Europe, in 6800 BC–4000 BC in Southeast Asia, and in about 2500 BC in Central America and Peru. Most of the regions that were developed for cultivating crops were the river basins with semiarid climate. Among them, some regions were densely populated. Once this was established, professionals began engaging in governmental affairs, war, religion, and the manufacturing industry.

The appearance, development, and proficient use of domestication resulted from the joint action of exterior environment and human intrinsic demand. The productivity released by cooking helped make the humans wiser, helped build relatively stable dwelling places, resulted in accumulation of plenty of food and tools for humanity, and resulted in a competitive difference between humankind and other animals. Humans began to have the potential to overcome difficulties and use energy of higher levels—they marched toward a more advanced civilization. The increase in population and demand for a higher standard of life that promoted humankind to develop energy of a higher level also advanced productivity to the next level—the domestication of animals and plants rose in response to the time and conditions.

Domestication was not accomplished in one stroke but needed to go through repeated selection and a long integration process. Dogs are models of domestication in the animal family. The dog was still a wolf feeding on meat 15,000 years ago when humans were a kind of omnivorous hunters. Both were hunters—competitors, each other's preys—enemies, or each other's food. How did they turn from being enemies to being friends? First of all, the growing gap between their strength ensured that they were no longer deadly foes that matched each other in strength. As humankind skillfully used fire and other hunting tools, wolves could not be considered as their rivals. The relationship between the two species gradually eased. At this time, humankind's teeth had become small, the stomach and intestines had shortened, and the requirement for food had become higher—humankind became picky and started to have leftovers. If food was rich, these leftovers were discarded as waste and became the food that the wolves got without any effort. Unintentionally, humans extended an olive branch to the wolves, and they reciprocated. The wolves had keen sensory abilities, such as a good sense of smell. They would roar when foreign invaders approached—the precursor of a dogs' furious barking—which became humankind's alarm and later; this is the dogs' ability to guard the entrance. The two enemies gradually formed an alliance. With a mutually beneficial and win–win alliance, they ate more and led a more secure life. Later, humans occasionally adopted wolf cubs, and the wolves became more and more like dogs.

Why were dogs distinguished from tigers, leopards, rhinos, and other animals that have not and cannot be domesticated by us so far? First, humans and wolves were not like tigers, leopards, rhinos, and other animals that lived alone; they lived in groups and had a strict hierarchical and organization structure as well as a good cooperative mechanism. Second, humans and wolves were very smart and could comprehend the benefit brought by mutual cooperation. Last, although the wolves were offensive, they were not as terrible as tigers and they gave humans the space to accept and change them. Beside dogs, horses, cattle, sheep, chickens, ducks, and geese were also gradually domesticated.

The domestication of plants was also a time-consuming practice and was fraught with many trials. Although not aggressive like animals, plants could neither speak nor act. Hence, there was no communication or interaction in the course of domesticating them. Domesticating plants started from their seeds. When there was surplus food, the seeds that were not eaten were stored or lost. Perhaps, the storing at that time merely involved piling the seeds in the fields; these seeds would take root and sprout in the coming spring and bear new seeds several months later. So, humankind did not need to gather seeds in the forests, and thus gradually learned farming.

Cooking helped humans to chew, digest, and absorb food; domesticating animals and plants made it convenient for humans to get food. From then on, humans occupied the top of the food chain, enjoying the most tasty delicacies and best nutrition, building stronger abodes and more beautiful villages, becoming more united and energetic, leading a better life, and giving birth to stronger and more intelligent descendants.

The Reindeer in Alaska

The course of domestication was usually intermittent and not successful every time. Domesticated creatures sometimes stopped spreading out. For example, the reindeer in Alaska was found in the northernmost part of the earth in the New Stone Age. They were a kind of deer with concave hooves suitable for walking on snow fields and could be found everywhere in the European and Asian continent and in North American tundra and boreal forest.

It is still not clear when the reindeer were domesticated. It is only known that it occurred long after humans domesticated cattle, horse, and other domestic animals. The reindeer provided meat and milk as well as materials for making clothes and tools in the farthest northern parts of the European and Asian continents. Besides, they pulled sleighs and wagons for humans. Since the humans who lived in these parts could not live on farming, the importance of the reindeer for them surpassed that of domesticating other animals.

When the last batch of Inuits* moved to northern America from Siberia, the reindeer still resisted domestication, although dogs had been domesticated. The confrontation between Inuits and the domesticated animals lasted many generations. Russia gradually annexed Alaska in the 18th and early 19th century, bringing weapons, fur trade, and civilization to the locals who were still in the New Stone Age then, but the reindeer was not included. The United States bought Alaska from Russia in 1867. Jackson,† who performed missionary work among the aborigines,

---

* Inuit (Eskimo), aborigines in the arctic regions, distributed around the Arctic Circle from Siberia and Alaska to Greenland.
† Sheldon Jackson (May 18, 1834 – May 2, 1909) was a Presbyterian minister, missionary, and political leader. During this career he travelled about one million miles (1.6 million km) and established more than one hundred missions and churches, mostly in the Western United States.

became the leader of the public service agencies in Alaska in 1885. In the hope of "civilizing" the Inuits, he began to teach them reading and arithmetic, and he raised money from individuals and the federal government for bringing in the reindeer from Siberia. He was convinced that just as cattle, horses, and sheep could bring benefit to the western states, domesticated reindeer could also bring actual benefits to Alaska.

One thousand two hundred and eighty reindeer were transported from Siberia to Alaska between 1891 and 1902, and there they were domesticated from then on. The population rapidly increased by one-third each year because they were fed well and were healthy. There were 600,000 domesticated reindeer in Alaska by 1932, and they were all the offspring of the first batch of domesticated reindeer. They were herded from the north coast of the Arctic Ocean to Kodiak Island and Aleutian Islands. But the number of the reindeer declined sharply after that. Only 200,000 were left by 1940, dwindling to 25,000, 10 years later, and these exist in only a few regions in the west of Alaska.

An increased rate of reproduction was the most important reason for the sharp decline of the reindeer. Six hundred thousand reindeer might find food in summer, but they could not survive in winter. The overgrazed pasture, which could be restored in several months or weeks in warm regions, took almost 30 years in the northern regions where it was freezing cold and where the sunshine duration was only several hours every day. The reindeer had to endure hunger after they ate up the reindeer moss.

The reindeer are not completely extinct in Alaska. Currently, there are thousands of reindeer on the Seward Peninsula and other regions, but they play an unimportant role in Alaskans' economic activities. "The large-scale industry which will perpetually develop and create a fortune" that Jackson had expected did not appear. Domestication could not be preserved. Sometimes, it was feasible, sometimes it was not.*

5.2 "Domestication" of Natural Forces

Humans had learned to make fire and had learned cooking and domestication of animals and plants. What did humans do next? The Chinese immortal poet Li Bai wrote at the end of his poem *Roads Are Difficult to Be Walked*: "Roads are difficult

---

* See *Human Energy History—Crisis and Hope*, (American) Alfred W. Crosby, China Youth Publishing Group, 2009.

to be walked, roads are difficult to be walked. There are many forked roads, which should I go? There will be the day when I sail across the sea by riding wind and cleaving waves." He also described in his poem *Early Morning Departure From Baidi Town:* "early morning departure from cloud-kissing Baidi, return to Jiangling thousand miles away in a day. The cry of the apes at the river banks does not stop, the light boat has passed ten thousand mountains." These lines fully exhibit the great power of wind energy and water energy, which impels human civilization to go faster and further.

Humans started seafaring activities in the New Stone Age. Domestication of wind energy allowed humankind to successfully sail across the sea; sailing boats (see Figure 5.1) rendered the most outstanding services in these seafaring activities. Chinese articles of that time have been found in Taiwan, Australasia, Ecuador, and other places. Pit Aas, a Greek seafarer in the 4th century BCE, sailed a boat from what is Marseille today to the mouth of Elbe by sea, becoming the earliest seafarer in the west. Warships that were 100 feet long were used in the sea warfare between Persia and Greece in 490 BCE. 15th Century CE was a period of great eastern and western seafaring development. During 1405–1433, Zhenghe, a Chinese seafarer, made 7 voyages to the western seas, passing through over 30 countries and regions, even to Somalia and Kenya in the eastern shore of Africa; in 1420, Portugal established a seafaring school; in 1487, Dias, a captain, sailed to the southern tip of Africa and named it Cape of Good Hope; in 1497, Vasco da Gama led a fleet to sail around Cape of Good Hope to India starting from Lisbon, Portugal. Later, Portuguese arrived in China and Japan. Columbus, an Italian seafarer, found the American continent in 1492. Seafaring promoted commodity trading and information exchange throughout the whole world.

We can use the monsoons to set sail on a voyage and return with the help of the wind blowing. But, water always flows from a higher place to a lower place in any season under the force of gravity since its density is about 1000 times that of air. We need to go against the stream after we go with it. How should we use water energy?

A waterwheel (see Figure 5.2) works by using the mechanical energy (potential energy and kinetic energy) of water flow to push its wheels or turbines and drive machinery for grinding floor, cutting wood, producing textiles, etc. It first appeared in Greece in the 3rd century BCE. Philo, a Greek, described it in the 2nd century BCE. It was used to produce cereals in the Han Dynasty (202 BCE–220 CE) in China. Heng Tan described in his book *New Theory*, written about 20 CE, that

Figure 5.1   600th Anniversary of Admiral Zheng He's voyage. (Courtesy of Stamps of Indonesia.)

Figure 5.2 Pedal waterwheel used by the government of Song Dynasty in China for spreading new agricultural technologies.

Fu Xi, one of three ancient emperors, invented pestle, mortar, and the watermill. Du Shi invented the shuipai (water-driven air blower), which is a kind of complicated mechanical device using water power. It drove machinery to make leather bags repeatedly open and shut so as to send air into iron-making furnaces.

Later, people learned how to build gutters and slope cannels to allow water to flow from a height to the top of a waterwheel and hit it, during which the energy of water and gravity, namely, the weight and impact of water flowing down from the top, were used. These well-made and well-rotating waterwheels could produce energy of over four or five horsepower. According to the French historian Fernand Braudel's (1902–1985)* research, on average, the cereals crushed by each watermill were five times more in volume than those crushed by two people operating a manual mill. In the early 12th century, there were 20,000 waterwheels used for

---

* Fernand Braude (1902–1985), a famous French historian. His main works include *The Mediterranean and Its Regions in the Period of Philip II*, *The Economic and Social History of France*, *The Material Civilization, Economy and Capitalism from 15th to 18th Century*, and *Capitalist Essays*.

crushing wheat, ores, etc. in France, which is equivalent to the power made by 500,000 laborers. From then on, waterwheels could be seen in places near rivers and streams all over the European and Asian continents, North Africa, and countries in the Atlantic and Pacific Ocean. They gradually spread all over the whole world. Some of them are still being used in many places, but the waterwheels used for generating electricity have been renamed as hydroelectric generators.

A windmill, (see Figure 5.3) which looks like the waterwheel, is a kind of motive power machine that converts wind energy into mechanical energy. It uses adjustable blades or stair rail wheels to collect wind power. A simple windmill consists of a wheel, which looks like the wildflower bitter fleabane, support, and transmission gear. The rotating speed and power of the wheel can be adjusted by changing the number of blades or the size of the blades in accordance with the amount of wind power. Windmills are very practical in the southeast of Iran, Chinese coastal regions, Holland, and other places (both in the past and at present). There were at least 8000 windmills in the damp Dutch countryside in 1650 where it was necessary to pump water. They were still standard windmills in American farms, especially on semiarid plains, in the late half of the 20th century. Windmills, especially wind power generators, can still be seen worldwide and are popular because they do not cause pollution to the environment.

Figure 5.3   Early windmill (in foreground). (Photo by Uberprutser.)

After the challenge of making fire and learning to use it, humans had to deal with the instability of water and wind. Water changes yearly and seasonally, and wind blows irregularly. Humans still had to find and domesticate more effective and stable forms of energy to promote civilization.

History of Windmills, Waterwheels, and Sails

Wind is the phenomenon of airflow in the natural world. Humans invented windmills in ancient times as dynamic machinery needing no fuel. As long as wind existed, they could convert its energy into kinetic energy through the rotation of their blades. Windmills were used for drawing water for irrigation and for grinding cereals in China, ancient Babylon, Persia, etc., more than 2000 years ago. Later, they became popular in Europe. Other than drawing water for irrigation and grinding cereals, windmills were also used for heating, refrigerating, shipping, and generating electricity. Holland was called the "state of windmills," and windmills became its symbol. Holland is situated in the westerly direction where prevailing winds blow from the west all the year round near the Atlantic. It is a country with a typical oceanic climate, where the sea winds and the land winds blow and never stop. Holland's lack of water and other power resources were compensated by its copious wind power.

There were about 1200 windmills in Holland at the end of the 18th century, each capable of producing 6000 horsepower. The largest one is as tall as a several-story building and has wings 20 m long. The Dutch love their windmills very much; so much so that the second Saturday of May each year is called "Day of Windmills," when all windmills in that country rotate at full speed. The windmills have also played a critical role in the project of reclaiming land from the sea to grow grain in Kinderdijk, Holland. Dutch windmills are famous all over the world. Holland has some of the largest windmills in the world; they have been listed as a world cultural heritage by the World Heritage Committee.

China's Yellow River waterwheels are comparable to Holland's windmills. They were also invented and used from early times. Bucket-type waterwheels (called "Persian dragons" in the west) were prevalent in 670 BCE, and dragon bone waterwheels were seen during 168–189. Yellow River waterwheels were ancient tools for lifting water for irrigation. Their shape was very much like the giant ancient carriage wheels, and their spokes were equipped with scrapers and water buckets. When water flowed through the blades, it pushed the waterwheel, making it rotate, and the buckets were filled with water, lifted up, tilted, and emptied one by one into the water channel heading to the farmland that needed irrigation. Lifting water for irrigation using the naturally flowing Yellow River water as the driving force resolved the difficulty of diverting water for irrigation to areas near the Yellow River with high banks and low water level.

The Yellow River waterwheels were characterized by the well-designed structure, well-chosen material, and these were highly efficient irrigation facilities and simple graceful works of art. For centuries, the Yellow River waterwheels rotated day and night at the banks of the torrential Yellow River. The people residing on the banks of the Yellow River love their waterwheels just like the Dutch love their windmills. Many poets composed verses and wrote articles in praise of the waterwheels in the Ming and Qing dynasties. Ye Libin, a poet in the period of Daoguang of Qing Dynasty, lyrically stated, waterwheels rotate repeatedly, empty milky water at river bends. I begin to believe Li Bai's wonderful verse, the water of Yellow River comes from heaven.

Another masterpiece whose history is longer than that of windmills and waterwheels, and that is used almost all over the world, is the sailboat. It uses both waterpower and wind power. A sailboat is a boat that uses sails as the driving force to move. It is a water vehicle with a history of over 5000 years, and it is more advanced than the boats and rafts, which simply use water power. There are many kinds of sailboats in the world. They can be divided into sloops, brigs, ketches, and barks based on mast number and as freighters, ferries, fishing boats, and war ships based on use. In the 15th century, Zheng He, a Chinese seafarer in the Ming Dynasty, leading a giant fleet, made 7 voyages to the western seas, reaching over 30 states in Asia and Africa; Columbus, leading a fleet, made many expeditions, and found the "new continent."

Windmills, waterwheels, and sailboats based on natural wind power and water power were outshone and almost forgotten with the start of the Industrial Revolution owing to the development of the steam engine, internal combustion engine, and turbine.* However, they gained new life through new innovation today in light of the energy crisis, ecological deterioration, and environmental pollution. Windmill generators are widely used in Holland, Xinjiang, Inner Mongolia, Tibet, China, and other places with rich wind power resources. New types of offshore and ocean-going motor sailboats whose sails are controlled by computers have been proactively studied in Japan, Britain, America, Brazil, and other countries since the 1970s.

Windmills, waterwheels, and sailboats are the greatest masterpieces which show that humankind can harmoniously get along with nature, as these are pollution free and do not lead to energy depletion. They have promoted human civilization and social progress. In 2006, the Yellow River waterwheels were included as protected Chinese intangible cultural heritage sites. In 2005, waterwheel and windmill stamps were jointly issued by China and Holland worldwide. All these actions commemorate the memory of the ancient inventions that use natural energy and demonstrate the willingness of people to harmoniously get along with nature.

---

* Turbine is a motor that uses liquid to impact and rotate blades to produce a driving force. Turbines can be divided into steam turbines, gas turbines, and water turbines. They are widely used in electricity generation, aviation, and navigation.

Chapter 6

Coal and Steam: Driving Force of the Industrial Revolution

6.1 Discovery and Early Use of Coal

In the early 18th century, although we had learned to use water power and wind power, most work, such as grinding cereals with stone mills or pedaling water-wheels, was still done manually and by using domesticated animals such as horses, cattle, and mules. Till the 18th century, plantation owners still met their production demand by using slaves. Even so, the demands of historic development for the driving force could not be met.

In the first half of the 18th century, the steam engine, which was run using coal, was invented. Technological revolution replaced individual manual production as the era of massive industrial production began. The first Industrial Revolution spread all over the European continent from England, to North America in the 19th century, and to all parts of the world later.

John Smeaton,* a forerunner of steam engine development, predicted that the total manual power produced only amounted to 90–100 W; it could not reach 500 watts even if everyone worked to maximum capacity.

It was then that coal appeared on the historical stage; until now a neglected fuel, it became a brand new natural force promoting technological innovation to leap at an unprecedented speed, bringing about political and economical reorganization

---

* John Smeaton (1724–1792) was an 18th century British inventor and mechanist who made many kinds of machine tools.

and integration, propelling population growth and migration, and improving the standard of life in many places. Almost everyone from central England to the Ruhr industrial region of Germany and from vast America to the Asian continent, except the gatherers and hunters who live in extremely remote areas and whose number continuously diminish, have been affected and have benefited from coal.

Coal is made up of carbon, hydrogen, oxygen, nitrogen, sulfur, and phosphorus, of which carbon, hydrogen, and oxygen make up about 95% of its organic matter. It is a solid flammable mineral resource formed from ancient plants that were buried underground and gradually went through complicated biochemistry and physical chemistry changes; the changes resulted in the formation of peat, lignite, soft coal, hard coal, and semi-hard coal. Plant remains under normal temperature and pressure of the earth surface were converted into peat or sapropel through peatification or saprofication.* This was then converted into lignite through diagenesis after it was buried and sunk deep underground since its base sank. Lignite was converted again into soft coal or hard coal through metamorphism when the temperature and pressure gradually increased.

Petroleum and natural gas, like coal, are also energy resources that nature has concentrated over millions of years for us. Petroleum comes from tiny phytoplankton that lived in the sea long ago, accumulated in waters without sufficient oxygen after they died, and were buried thousands of meters underground. The phytoplankton were converted into liquid petroleum by pressure and high temperature. Petroleum has a much higher energy density than coal and is more convenient for storage and transportation. Natural gas is a kind of multicomponent fixed gaseous fossil fuel, mainly consisting of alkanes—mostly methane with a bit of ethane, propane, and butane. It mainly exists in oil fields and natural gas fields; small quantities can also be found in coal beds.

Coal, petroleum, and natural gas were all formed directly or indirectly from plants that had gathered solar energy through photosynthesis for a long period of time. Large amounts of plant matter could only form a small amount of fossil fuel. For example, 90 tons of plants, equivalent to 40 acres (about 161,900 m²) of wheat, could only yield 1 gallon (about 3.79 L) of petroleum. Fortunately, the earth has been regularly storing plants into the energy treasury to help later human generations to create industrial civilization.

Coal and the steam engine were the perfect combination to replace firewood and became the primary driving force. It was widely used in mining, grinding flour, making paper, manufacturing textiles, smelting, etc. Manual production was replaced by machines, which promoted rapid development of industry and led to prosperity. This was another great step in the history of human energy after drilling wood to make fire.

---

* Peatification refers to the process in which plant remains accumulated in swamps and are converted into peat. Saprofication is the process during which the remains of lower order creatures in swamps are converted into sapropel, a kind of slushy matter containing water and asphalt.

Marco Polo's "Black Stone"

Marco Polo was a famous Italian traveler of the 13th century. He traveled to many provinces and cities of the Yuan Dynasty, China, and wrote the widely popular book *The Travels of Marco Polo*. In the book, he stated "China's fuel is neither wood nor grass but a kind of black stone."

China was the first to use coal as an energy resource. There is a record of its use in the ancient geographic literature *Shan Hai Jing* (*A Book of Mountains and Seas*). Coal is also called graphite, stone carbon, black firewood, black gold, and flammable stone. Burned coal cinder and unburnt coal cakes have been found in brick beds in the houses of Fushun, which is located in the northeast of China, and the iron-making sites in the middle of China. It shows that coal was universally used for heating and iron making in ancient China.

According to another ancient geological book *Shui Jing Zhu* (*Commentary on the Water Classic*), the Ice Wall Terrace Coal Mine built by Cao Cao at Ye County (now in the west of Linzhang County, Henan Province, China) in 210 CE in the period of the Three Kingdoms was 50 m deep and stored thousands of tons of coal. Coal mining underwent a relatively rapid development in the Song Dynasty. Huge amounts of coal were found, coal mining organizations were specially established, and a coal monopoly system was followed by the government. Millions of people in Bianliang (now Kaifeng, Henan Province), then the capital of China, depended on coal instead of firewood. Coal-mining sites at Hebi and Henan province show that the mining industry was technically very sophisticated, and even the facilities were more comprehensive.

The book *Tian Gong Kai Wu* (*Exploitation of the Works of Nature*) was written during the Ming Dynasty. In this book it describes how Chinese people cope with the mashgas generated during coal mining process at that time. Please modified if the sentence has flaw: one end of one thick hollow bamboo pole was first sharpened and then inserted into coal beds to release gas out of the coal mines, which was very ingenious.

The Chinese used coal during spring and autumn and Warring States Period, more than 2000 years ago. It entered common people's houses as an important fuel at the end of the East Han Dynasty, but was not used in Europe until the 16th century. It is no wonder then that Marco Polo did not know what it was.

6.2 Steam Engine and the First Industrial Revolution

Britain is the birthplace of the Industrial Revolution. It is not completely clear why the Industrial Revolution first took place in Britain, but the abundant coal reserves in the country could be one of the key factors. At the same time, insufficient forest resources stimulated the British to use coal as the main energy source instead of biomass. The timber price in Britain increased sevenfold during the period from 1500 to 1630, much faster than inflation. The tree census in 1608 showed that there were 232,011 trees altogether in the seven forests in Britain; it reduced to 51,500 trees in 1783. Thus, people had no other choice but to mine more coal.

Coal, with its high energy content was not only used for cooking but also provided a driving force to the heat engine, a kind of motive power machine that converted the chemical energy of fuel into mechanical energy. Its earliest representative was the steam engine that came into being in the 18th century. People saw the lids of their kettles lifted up by steam when they were heating water—thus, they found the power of steam. Hero of Alexandria, in the 1st century, even made a rotating lawn sprinkler that was driven by steam. In 1654, Otto von Guericke, a German, used two large copper hemispheres to demonstrate the power of atmospheric pressure. When the rims were sealed and the air was pumped out to create a vacuum, the sphere could not be pulled apart by teams of sixteen horses. Christiaan Huygens, a Dutchman, believed that if gunpowder was exploded under a piston in a metal cylinder, it could raise the piston up to the top of the cylinder, expel the air out of the cylinder, thus filling the cylinder partially with vacuum. Then the piston would be pushed down into the vacuum by atmospheric pressure. Inspired by this, people thought of the working theory. Thomas Newcomen* invented the steam engine. He was an excellent artisan, trained in the background of western Europe's economic growth. Although not educated, his biographer regarded him as "the first great mechanical engineer in history."

Newcomen set up a steam engine close to a coal mine 46 m deep in Staffordshire, Britain, in 1712. Its boiler could hold 673 gallons (about 3060 L) of water. The cylinder was 2.4 m high and 53 cm in diameter; the clearance between the cylinder and piston was filled with wet leather. The boiler was heated with coal to produce steam that pushed the piston upward, and then cold water was injected into the cylinder to condense the steam into liquid to fill the cylinder with vacuum. Finally, atmospheric pressure pushed the piston into this cylinder. Power engineers came into existence from then on. The motion of the piston was achieved by the push of a connecting rod that was connected with it. The first steam engine had a speed of 12 strokes†/min, which is equivalent to 5.5 horsepower (4.1 kW), and had the power to lift 10 gallons (about

---

* Thomas Newcomen (1663–1729), a British engineer, was one of the inventors of the steam engine. The atmospheric steam engine invented by him was the predecessor of Watt's steam engine.
† Reciprocating motion, used in reciprocating engines and other mechanisms, is back-and-forth motion. Each cycle of reciprocation consists of two opposite motions: there is a motion in one direction, and then a motion back in the opposite direction. Each of these is called a stroke.

45.5 L) of water. This caused a great sensation in Britain and Europe, which lacked mechanical power at that time. At least 1500 Newcomen steam engines were made worldwide in the 18th century, which is enough to show how great the social demand then was. When Newcomen died, his invention had appeared in Saxony, France, and Belgium. By 1753, the first-generation Newcomen steam engine in the U.S. was seen in North Arlington, NJ.

Newcomen's invention was the first machine that could provide huge amounts of mechanical power which was not physical strength, water power, or wind power but a new natural force. It used coal to heat water and produce steam to do work. It was also the first machine that used pistons in cylinders and could run day and night. Without the timely appearance of Newcomen's steam engines, the 18th century British coal industry would have perished or remained stagnant, and England would not have been industrialized.

Watt* was the first outstanding figure among all new-generation engineers; he was well-educated and had a good relationship with scientists and capitalists. He improved Newcomen's steam engine in 1764. Rather than ejecting cold water onto a hot cylinder, he thought to use steam power and the falling piston to push the steam into the unheated space for condensation. For over 10 years, Watt made two kinds of steam engines, with satisfactory operational results. By 1800, the driving force produced by the Watt steam engine (see Figure 6.1) per unit weight coal was 3 times that produced by the latest Newcomen steam engine. Its application scope was enlarged from drawing water to grinding flour, making paper, smelting, etc. But the Watt steam engine was still consuming too much coal and was too huge and cumbersome to serve as a driving force for vehicles and ships. Besides, it was similar to Newcomen's steam engine that needed to rely on atmospheric pressure to push the piston into the cylinder; therefore, its driving force and speed were limited.

At the end of the 18th century and in the early 19th century, after Watt's patent expired, different methods of improving and applying technologies of steam engines sprang up. High-pressure steam engines were invented within a few years. These used steam to pull pistons out and push them into cylinders; the crank was still connected with the connecting rod. It greatly enhanced British industrial production efficiency and spread worldwide several years later. In 1800, Richard Trevithick designed the high-pressure steam engine that could be installed on larger vehicles. In 1803, he used it to drive the locomotive on a circular track and found people who liked novelties to ride and charged them. That is the rudiment of locomotives. George Stephenson continuously improved the locomotive and created the "rocket" steam locomotive

---

* James Watt FRS FRSE (30 January 1736 (19 January 1736 OS) – 25 August 1819[1]) was a Scottish inventor, mechanical engineer, and chemist who improved on Thomas Newcomen's 1712 Newcomen steam engine with his Watt steam engine in 1781, which was fundamental to the changes brought by the Industrial Revolution in both his native Great Britain and the rest of the world. (From Wikipedia)

Figure 6.1   Model of the Watt steam engine. (Photo by Nicolás Pérez.)

in 1829. With a carriage accommodating 30 passengers and a speed of 46 km/h, it caught every country's attention, and opened the railway epoch.

The Steam Engine Changed the World

The textile industry was most affected by the steam engine; it was the first industry to become mechanized. The output per unit time of a spinning machine driven by a steam engine was equivalent to the total output of 200–300 manual spinners in 1800.

From then on, the spinning mills in Britain and New England* produced a large quantity of cotton yarns and factories produced garments each year that were cheaper and better than previous products. The success of using the steam engine in the textile industry made southern America think of using it in single cotton planting areas. This labor-intensive farming still relied on slaves, reviving the gradually perishing concept of slavery.

The Industrial Revolution of the 19th century quickly changed global economy; it also redistributed the world's power. For instance, British factories almost wiped out India's

---

* New England is the region including six American states, situated in the northeast part of the American continent, bordering on the Atlantic, and adjoining Canada.

traditional textile industry by their speed and output, resulting in the loss of livelihood of thousands of Indian peasants. In the 18th century, Indian, Chinese, and European GDP accounted for 70% of the world's total GDP, and the GDP of each of the three accounted for about one-fourth of the world's total GDP. By 1900, China's GDP fell to 7% of the global finished products ratio and India's slumped even more to 2%, but Europe's rose to 60%, and America's went up to 20%.

The steam engine had an extremely important influence on the global transportation industry. The early steam locomotive (see Figure 6.2) began to run with a roaring sound in the early 19th century. The locomotive named "Rocket," along with a carriage, arrived at Manchester from Liverpool in 1830. Ten years later, the British railway system extended 2253 km, the European railway system 2414 km, and the American railway system 7403 km. America built the first transcontinental railroad in 1869. All parts of the world made many astonishing plans for railroad construction, such as the railroad from Cape Town to Cairo, the Trans-Siberian railway.

The steam engine also brought about a great change to seafaring. In 1838, the *Sirius* and the *Great Western* left the British ports for New York to vie for the honor of being the first

Figure 6.2   Steam locomotive.

steam-powered vessel that finished an ocean-going voyage. The *Sirius* won. It took 18 days and 10 h to reach the destination at a speed of 6.7 knots/h (about 12.4 km/h) on average. The *Great Western*, which had four boilers and the two latest engines, took 15 days at a speed of 8 knots/h (about 14.8 km/h) on average and consumed 200 tons of coal. The new ship saved about half the time than a ship sailing across the Atlantic only depending on wind power.

Steam power not only enhanced the speed and reliability of goods transportation but also promoted population migration. People traveled from the countryside to cities and from developed regions to undeveloped regions by train. A large number of Europeans immigrated to their overseas colonies and to America by steamships. At the same time, millions of people left India and China for America, South Africa, East Africa, Mauritius, Pacific Islands and other parts of the world to sell their labor to plant crops and build docks, highways, and railways. The number of the total immigrants in the world in 1830–1914 reached one hundred million. Most of them traveled far away across the sea by steamships no matter where they came from and where they were going.

With the advent of the steam engine, productivity increased, and global power and influence reshuffled. The global hegemony center was transferred, and population moved frequently. The American politician Daniel Webster[*] praised the steam engine: "It can sail, draw water, dig, carry things, drag, lift, hammer, spin, print. It is similar to a man more like an artisan. Stop physical work, use your skill and wisdom to control it, it will take over all your hardships. There is no longer any one who feels tired, needs rest, gasp for breath. We can't predict how we will improve and use this kind of power in the future. Any guess will turn out to be futile."

In the past, driving force merely referred to physical strength. To use and abuse slaves and serfs was no doubt the most effective way to use this kind of power. But, the best way of obtaining limitless driving force was to have a steam engine.[†]

---

[*] Daniel Webster (1782–1852), a famous American politician, jurist, and lawyer, was American Secretary of State three times and senator for a long time.
[†] See Alfred W. Crosby, *The History of Human Energy—Crisis and Hope*, China Youth Press, 2009.

Chapter 7

The Light of Electromagnetic Measurement in the Information Civilization

7.1 Advent of the Internal Combustion Engine Era

After 1870, science and technology developed by leaps and bounds. All kinds of new technologies and inventions emerged in an endless stream and were used for industrial production, thereby greatly promoting economic development. The invention and application of new things no longer took millions of years as in ancient times.

As people made efforts to improve steam engines, their efficiency was continuously enhanced. The operational efficiency of the powered steam engine of the 1900s produced power five times that of the steam engine made in 1830 as paddle wheels were replaced by screw propellers. However, this improvement was not enough. A fuel whose energy density was higher than that of coal and that was easier to transport had to be found to realize a real breakthrough. People looked for more efficient energy forms. The beneficial match of petroleum with the combustion engine generated more efficient energy and quickly captured people's favor.

Petroleum, also called crude oil, is a kind of thick dark brown liquid and is found in the upper crust of the earth. Similar to coal, it is a fossil fuel formed from the remains of animals buried under rock and subjected to intense heat and pressure for millions of years. The energy density of petroleum is about 50% higher than that of coal; it is present in liquid state and is easier to pack, store, and transport.

In the 19th century, the petroleum industry developed slowly, and refined petroleum was mainly used as a fuel for kerosene lamps. In the early 20th century, with the invention of the internal combustion engine, the situation changed greatly. Since then, petroleum has always been the most important fuel for the internal combustion engine.

The revolution of transportation had increased the demand for fossil energy. The horse, though practical, tires easily, is sometimes not cooperative, and cannot be repaired if it is "broken." In the middle of the 19th century, the steam engine replaced horses in carrying large number of passengers and huge cumbersome goods between cities, big and small. But the construction cost of railways and large-scale steam ships were expensive, and engines and steam boilers were too cumbersome to enter daily life, which limited people's traffic demand for excursions.

Engineers and inventors began to think of new ways of transportation. They knew that energy coming from the sun and coal could heat water into steam that could be used to drive pistons. This burning of fuel was indirect and inefficient; so, why not burn the fuel inside the piston? Would not an "internal" combustion engine be better than the steam engine? These thoughts also inspired Newcomen and other steam engineer researchers. In the 17th century, some inventors, inspired by the gun principle, tried to make a piston driven by gunpowder but failed because they could not control the speed of explosion nor could they continue to fill gunpowder into the cylinder after explosion. But the concept of the internal combustion was spread wide and far.

Nicolaus August Otto,* originally a salesman, became an engineer in 1863 and built a machine that reminded people of Newcomen's steam engine, but its fuel could be a gas that burned inside the cylinder to push the piston upward. It was then pushed down into the cylinder by its own weight and atmospheric pressure. Although the machine brought fame for him, it was not as powerful and smooth as he imagined.

Otto spent another 10 years and finally made a machine that could carry out a series of combustions that could be controlled. He applied for a patent for the epoch-making four-stroke engine, and he named it the "Otto engine" in 1876. The engine started with batteries or a crank handle and pushed the piston upward to let fuel and air enter the cylinder. The piston was then pushed down into the cylinder to let the fuel and air mix to ensure complete and even combustion of the fuel, which then pushed the piston upward again—the power stroke. The piston was pushed down into the cylinder again in the final stroke to expel the exhaust after combustion. In this way, the combustion automatically runs by inertia, and theoretically would not stop until the fuel runs out. The Otto engine was the first practical internal combustion machine; it is still used in our cars now.

---

* Nicolaus August Otto (1832–1891), a German scientist, invented the practical four-stroke cycle internal combustion engine.

Figure 7.1 Early plane.

The engine invented by Rudolf Diesel* was the most famous internal combustion engine. It used crude oil as fuel and ran more efficiently than other engines. Before long, a more highly efficient internal combustion engine—the turbine—came out.

The invention of the internal combustion engine gave us the key to the sky (see Figure 7.1). The French inventor Clément Agnès Ader† successfully invented the first plane. His plane III was an aircraft similar to the two wings of a bat, equipped with two steam engines of 20 horsepower (about 15 kW). It took off successfully in 1897. But the steam boiler full of water was a huge burden for the plane. The Wright brothers‡ designed a unique internal combustion engine that could produce 12 horsepower (nearly 9 kW); the total weight of the motor itself plus fuel and fittings was only 200 pounds (about 90.7 kg). One plane equipped with this motor flew into the blue sky in 1903.

The airplane gave wings to mankind and made their dream of flying in the sky come true. It made science and technology seem almost magical. However, another invention—the car—was more popular and practical. The three-wheel cars made by Mercedes Benz, Daimler, and Maybach in 1885 were the first cars that could run on the road at that time. The first commercial Mercedes Benz car formally ran on the road 5 years later. At the beginning, the car was regarded as a substitute for

---

* Rudolf Diesel (1858–1913), a German engineer and inventor, invented the diesel engine.
† ClémentAgnès Ader (1841–1925), a French engineer, invented the first plane.
‡ The brothers, Wilbur Wright (1867–1912) and Orville Wright (1871–1948), were American inventors of the airplane.

the carriage and a plaything for the rich. The Americans, represented by Henry Ford,* produced the Model A in 1903 and the Model T 5 years later after trying 20 different car models. Later, based on the chassis of this type, nine types of cars were produced, including two-seat cars and light trucks. The Model T was sturdy, durable, and easy to operate and repair. So it became the cheaper means of transportation suitable for individuals and families. In 1925, the Model T cost 260 US$ and could be afforded by everyone from the middle class to the common worker of Ford Company. Accumulatively, 16 million were produced before its production stopped in 1928. As the car industry developed, the number of each country's cars increased with each passing day. By the end of the 20th century, the number of horses used for transportation had reduced, but cars, excluding trucks, buses, tractors, tanks, and other vehicles, increased to 500 million.

People are heavily dependent on and even greedy for petroleum. But the petroleum reserves left to us by nature are limited and will soon be overdrawn. Petroleum no longer merely brings us civilization but also wars. Humans have imperceptibly become "prey" and "sacrificial offerings" to energy while trying to domesticate it.

Blood of Modern Industries—Petroleum

Petroleum in the form of natural pitch was known in ancient Egypt, Babylon, and India as early as before the 10th century BCE. Petroleum that oozed out of the surface was stored after long-term steaming and used for construction, prevention of corrosion, binding, decoration, and as a pharmaceutical product. The ancient Egyptians could even calculate the quantity of oil from seepage. There are also records related to collecting natural petroleum in the cuneiform† found in the coastal regions around the Dead Sea.

It was the Europeans who first extracted kerosene from crude oil and used it for lighting. A pharmacist of Lvov made a kerosene lamp with the aid of a blacksmith during the period 1840s–1850s. In 1854, the kerosene for lamps became a

---

* Henry Ford, an American car engineer, entrepreneur, and founder of Ford Motor Company, first produced cars using a streamlined mass production technique, making cars affordable to the general public.

† Cuneiform was the script used in western Asia in ancient times. It was written on stones and clay tablets. Manually dug petroleum wells appeared around Susa, Persian Empire, in the 5th century. It was also the Middle East countries that were the first to use petroleum in wars. Homer described in Iliad: "Trojans ceaselessly cast flame onto the ships, which produced flame difficult to be put out." When Persian king Seleus prepared to usurp Babylon, someone warned him that the Babylonians were adept at street battle. Seleus replied that he would attack with fire: "We have a lot of pitch and pieces of flax, fire can be soon cast about, those who are on roofs will be engulfed by fire if they won't escape quickly."

commodity on the Vienna market. In 1859, Europe extracted 36,000 barrels of crude oil, mainly from Galicia and Romania.

At present, petroleum has become one of the main energy resources in the world and occupies an extremely important position in the global economy. Petroleum is the raw material of high-quality power fuels. Cars, diesel locomotives, airplanes, ships, and other modern vehicles use its derivatives—gasoline and diesel—as power fuel. Also supersonic airplanes use the high-quality fuel extracted from petroleum as fuel. Petroleum is also the raw material for refining high-quality lubricating oil. All lubricating oil that is applied to every rotating mechanical "joint" is a petroleum product.

Petroleum is still an important chemical material from which over 5000 kinds of organic synthetic materials are made. For example, colorful and durable synthetic fibers such as dacron, nylon, acrylic fiber, and polypropylene fibers, synthetic rubber comparable to natural rubber, aniline dyestuff, washing powder, saccharin, man-made leather, fertilizer and dynamite are all made using petroleum.

Petroleum can also be made into a synthetic protein through microbial fermentation. A kind of wax-chewing bacteria will multiply at astonishing speed after they are put into petroleum, and a kilogram of them contain protein equivalent to that of 20 eggs. If half of the wax contained in the over 3 billion tons of petroleum produced each year in the world are made into protein, we will get 150 million tons of protein, which is a very significant resource. The wax-chewing bacteria are used as feed at present and will be made into delicious nutritious food and be served on our tables in the near future.

The petroleum coke and pitch left behind after petroleum is refined are also precious. The petroleum coke can be used as electrodes to improve steel output and as the raw material for making graphite; pitch can be used for making felt paper or for building roads. No wonder, petroleum, with such rich and important applications, is praised as "the blood of modern industries." It is worthy of the name.

7.2 Change from Magnetism to Electricity

Electricity is a kind of natural phenomenon, and energy and is caused by the movement of electrons. Lightning initiated human civilization by helping us get fire. We learned to make fire very early on but did not domesticate electricity until the second Industrial Revolution—maybe because it seemed more mystic and dangerous, and its laws more difficult to master. Electrical energy is satisfying. It can be transmitted in great amounts rapidly and can be converted into thermal energy, mechanical energy, chemical energy, and optical energy.

Although electricity was domesticated only recently, its history is long. Ancient Greeks found when amber was rubbed, it could attract feathers. So the word "electricity" came from Greek "ηλεκτρου," which means amber. It was not until William Gilbert,* a doctor for Queen Elizabeth I,† found that in addition to amber, many other objects also produced magnetism after they were rubbed that people began to study electricity. One hundred years later, Francis Hoxby, another British man, invented an electrostatic device that could be put into practice. He installed a crank on a hollow glass ball. The glass ball could produce sparks for entertainment and could also generate sufficient electricity for experiments when it was rotated at high speed and rubbed against a leather cushion.

Pieter van Musschenbroek‡ invented an electricity-storing device in Leiden, Holland, called the "Leiden jar." At the beginning, it was a simple water jar, in which half a piece of metal wire was inserted. Later, it was wrapped both inside and outside with metal. The electricity could be kept for a while and even several days after it was conducted from a large-scale Hoxby electrostatic device to the metal wire inserted in the jar.

The Hoxby electrostatic device and the Leiden jar generated little electricity and so could not meet the demand of important scientific experiments. Alessandro Giuseppe Antonio Anastasio Volta,§ an Italian, invented a stable electric device in the 18th century. He made the so-called "Volta electric pile" with sheet copper, sheet zinc, and wet cardboard after it was dipped into brine. The sheet copper released electrons to the wet cardboard, then the sheet zinc received them, and the unfixed electrons flowed out through the single metal wire. Thus, the earliest battery was formed—it could provide current until the liquid completely evaporated or entered into the zinc.

The British scientist Michael Faraday¶ was another hero who made important achievements in electromagnetism. The Dutch scientist Hans Oersted** discovered the magnetic effect of current in 1820. He put a piece of very fine platinum wire on the

---

* William Gilbert (1540–1605), physicist of The Royal Society, made important contributions in the field of electricity and magnetomechanics.
† Elizabeth I (1533–1603) (reign 1558–1603), queen of the Kingdom of England and Ireland, was the last monarch of the Tudor Dynasty. She made England one of the strongest nations in Europe during her rule, which lasted for nearly half a century. English culture reached its peak during this period.
‡ Pieter van Musschenbroek (1692–1761) was a Dutch scientist.
§ Alessandro Giuseppe Antonio Anastasio Volta (1745–1827), an Italian physicist, invented the "Volta electric pile" in 1800.
¶ Michael Faraday (1791–1867), a British physicist and chemist, invented generators and electromotors.
** Hans Ørsted (1777–1851), a Danish physicist, chemist, and man of letters, found the current magnetic effect.

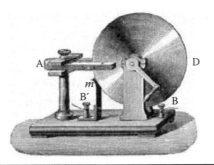

Figure 7.2   The disc-type generator designed by Faraday.

top of a glass-topped small magnetic needle. The small needle deflected the moment power was switched on. He found, through repeated experiments, that the magnetic needle would deflect as long as it was near current. Its deflection directions at the top and bottom of the wire were reverse, and its deflection was not affected if nonmagnetic objects such as wood, glass, water, or resin were put between the conductor and the magnetic needle. Faraday continuously repeated and improved Oersted's experiments and made both the hand and sheet copper and sheet zinc rotatable. He wondered if there were other motion modes, whether electricity and magnetism could "move." He wrote a note to himself in 1822: "to convert magnetism into electricity."

Nine years later, Faraday found that the galvanometer showed the presence of electricity when the magnetic rod went through the metal coil of the galvanometer. This proved that magnetic motion could generate electricity. Based on this, he made the first "generator" (see Figure 7.2). Weak current was produced by turning the sheet zinc between the positive and negative electrodes of a horseshoe magnet connected with the crank. (The American Henry Joseph* discovered the same phenomenon almost at the same time but published it later than Faraday.)

Both Faraday and Ørsted found the close link between electricity, magnetism, and motion. Motion could be produced as long as the first two items were controlled; electricity could be produced if the last two items were controlled. The first generation generators after Faraday were similar in design. They all used a metal wire loop rotating between the two electrodes of a permanent magnet or a permanent magnet rotating between the ends of a metal wire loop. Faraday's generator could only generate weak electric current, and the early generators designed after him using natural magnets were not enough for industrial use, but they could produce more powerful electricity.

Joseph Henry and other inventors wound current-transmitting insulated metal around big horseshoe type soft iron cores to convert them into electromagnets. Electric power could be adjusted in accordance with different requirements.

---

* Joseph Henry (1797–1878), an American physicist, made outstanding contributions to the field of electricity.

The device designed by Henry with electromagnets could lift 1 ton at one go. In 1866, when the German Werner von Siemens* experimented with an electromagnet in the generator he attempting to make, electric current ran out wildly.

Faraday produced electricity by manually rotating a coil loop between the two poles of a magnet. In this sense, the power source of the first electricity generation was physical strength. This generator converted the mechanical energy of the internal combustion engine, waterwheel, and windmill into electric energy, which was then converted into kinetic energy on the electromotor. The energy was used for rotating equipment such as transportation and spinning machines.

Electricity brought power and light. Arc lights were the earliest widely used electric lighting equipment. By first connecting two carbon rods to produce strong current and then disconnecting them to form electricity which burned the carbon rods, brilliant, bright light was generated. Arc lights are the best lighting equipment for big spaces such as urban squares and gymnasiums; they could also provide lighting for streets and large blocks, but were not fit for house lighting because they were too bright and made an unbearable buzzing sound.

A lot of people began to develop the incandescent lamp for providing ideal house lighting. A great amount of data accumulated, which greatly benefited later inventors. Thomas Alva Edison† and his workers tried hundreds of kinds of materials and finally developed the vacuum bulb with a carbonized bamboo filament. He applied for a patent for it in 1880. By the 20th century, wolfram became the best choice for filament, and the bulb was filled with inert gases instead of vacuum.

The greater advantage of electricity besides its wide application is its extremely easy transmission. Generators could be directly installed at coal mines or near Niagara Falls from where the electricity generated by coal or water was transmitted to remote users through cables.

At the end of the 19th century, western civilization quickly entered the electrical epoch; other regions followed and started out on the electrical road.

Entering the "Information Era"

Information exchange is very crucial in social activities. In the past, production of goods was extremely inefficient, science and technology was in the nascent stage, and a means of long-distance communication was undeveloped. Nowadays, long distance, and even global, information exchange can be completed in a moment with various communication approaches.

---

* Werner von Sviemens (1816–1892) was a German engineer scientist, entrepreneur, and inventor of the electromotor, generator, tramcar, and compass-type telegraph.
† Thomas Alva Edison (1847–1931), an American inventor and entrepreneur, had over 2000 inventions, including phonograph, motion picture cameras, and incandescent lamps. He set up the General Electric Company.

The telephone, invented by Alexander Graham Bell* on March 10, 1876, is said to be the most convenient and practical one.

Alexander Graham Bell was born in Britain in 1847. From childhood, he loved finding solutions to problems; he disassembled and assembled all kinds of articles. At the age of 15, he improved the traditional watermill and increased its production efficiency greatly. Later, he emigrated to Canada and America and invented the apparatus that could help deaf-mutes hear; he also improved phonographs.

Prior to Bell, ideas about telephones started to develop with the invention of the telegraph in the 19th century. The train and ship had become the tools of spreading information, but long-distance, and even transoceanic, communication was time-consuming. As a consequence, a kind of simple and fast means of communication was longed for and implemented in the early 18th century.

The American Samuel Finley Breese Morse[†] (see Figure 7.3) began to develop the wired telegraph (see Figure 7.4) using the electromagnetic principle in 1832. He succeeded in 1837 and wrote the telegraph code for English letters using different combinations of long and short pulse signals. In 1844, he set up an experimental telegraph line between Washington and Baltimore, which was a success, and this was when the wired telegraph was formally introduced.

In 1875, Bell thought, after careful observation of Boston's telegraphs, that the key for the mutual exchange between current signals and mechanical motion was an electromagnet field. Inspired by this, Bell began to design and make an electromagnetic telephone and finally succeeded after countless explorations and failures. Telephones (see Figures 7.5 and 7.6) came into being in June 1876 and were soon popularized worldwide.

Telephones greatly expanded our sensory function and promoted communication to a revolutionary era. Thus, Bell made a great contribution to social progress.

As communication technology continuously developed, radio technology appeared at the end of the 19th century and in the early 20th century.

---

* Alexander Graham Bell (1847–1922), an American inventor and entrepreneur, obtained the patent for the first practical telephone and established the Bell Telephone Company (predecessor of AT&T).

[†] Samuel Finley Breese Morse (April 27, 1791–April 2, 1872) was an American painter and inventor. After having established his reputation as a portrait painter, in his middle age Morse contributed to the invention of a single-wire telegraph system based on European telegraphs. He was a co-developer of Morse code and helped to develop the commercial use of telegraphy.

Figure 7.3   Father of the wireless telegraph—Morse.

Heinrich Rudolf Hertz,* a German physicist, identified the electromagnetic wave phenomenon through experiments. The electromagnetic wave was called the Hertz wave in the scientific community. Using the Hertz wave to transmit information, a French scientist Édouard Branley, a British scientist Oliver Joseph Lodge, a Russian scientist Popov, and other people carried out all kinds of experiments, laying the foundations for the invention of the radio.

Hertz wave research also attracted the attention of the Italian inventor Guglielmo Marchese Marconi†. Targeting the difficulties of setting up wired tele-

---

* Heinrich Rudolf Hertz (22 February 1857–1 January 1894) was a German physicist who first conclusively proved the existence of the electromagnetic waves theorized by James Clerk Maxwell's electromagnetic theory of light. The unit of frequency–cycle per second–was named the "hertz" in his honor.
† Guglielmo Marchese Marconi (1874–1937), an Italian electrical engineer, was the inventor of radio technology.

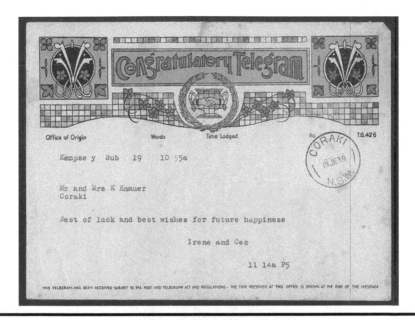

Figure 7.4   The telegraph made long-distance information transmission possible.

Figure 7.5   Early artificial telephone exchange.

phones and telegraphs, he put forward a bold assumption: could Hertz waves be used for long distance communication? With this question in mind, he began to broadly collect data about Hertz waves and telegraph communication and carefully research various scientists' uses of electromagnetic waves for communication to conduct experiments of radiotelegraphs. He finally succeeded in the experiment

Figure 7.6   Magnetic telephone. (Photo by Dori.)

of radiowave transmission through continuously improving his equipment in 1895. He succeeded again in communicating over 12 km with the support of the British. He gained the patent for the invention of the radio in 1896 in Britain and successfully carried out the first transoceanic radio communication on the British western coast in 1897. He conducted successful experiments in radio transoceanic communication over 2000 miles from Cornwall, Britain, to Newfoundland, Canada, in December 1901, marking wireless telegraph as a global undertaking. Later, Marconi further improved wireless telegraph devices, developed the horizontal directional antenna, and used rectifier tubes in radio communication devices.

Radio technology made long distance communication a reality, which made information exchange between different parts of the world much more convenient than before; furthermore, it also pressed ahead the course of entry into the information era.

7.3 Nuclear Power Developed in Controversy

The fossil fuels coal, petroleum, and natural gas, which were formed in the earth a long ago, have been depleted by industrial use leading to an international "energy crisis." Efforts must be made to find new energy to solve this problem. Nuclear technology, in which important progress has been made after many setbacks and that have been recognized internationally, will hopefully become an important means of solving the energy crisis. It is an important result in science and technology with a history of only about a century.

Although the nucleus is very small in size, it contains large amounts of energy. The energy (E) released by nuclear fission or fusion equals the product of mass (M) multiplied by the square of light speed (C) in accordance with Einstein's mass–energy equation. A uranium nucleus will split into two new nuclei with the simultaneous release of energy when a hot neutron bombards it—this is called nuclear fission. When it splits, a uranium nucleus can simultaneously release two or three fast neutrons, which can become hot neutrons when they pass through a moderator to reduce their speed; another uranium nucleus will split if one of the hot neutrons bombards it. This kind of continuous fission reaction is called a chain fission reaction. Large amounts of uranium nuclei can continuously fission, with continuous release of great amounts of energy in a short time because the speed of a fission reaction is so fast that 1000 generations of neutrons can be produced in a second.

Nuclear power generation is an important form of energy used at the present. Its main principle is to use the heat produced by uranium fuel in its nuclear fission chain reaction to heat water into steam with high temperature and high pressure that can drive steam engines and generators. The heat released by nuclear reaction per unit is millions of times greater than that of fossil fuel combustion. The Soviet Union built the world's first nuclear power plant—Obninsk Nuclear Power Plant (with an installed capacity 5 MW)—in 1954; following the Soviet Union, Britain, America, and other nations built nuclear power plants one after the other. China has 17 nuclear power units with 14.76 million kW installed capacity in grid-tied operation. In addition, nuclear energy is still used in military applications such as atomic bombs, nuclear-powered aircraft carriers, nuclear-powered submarines, etc.

Nuclear power generation is an effective measure for solving the world energy crisis, and it has many advantages. First, there are rich nuclear fuel resources such as uranium, thorium, deuterium, lithium, boron, and uranium fuel supply that can meet a longer time demand after 2020 when the fourth-generation nuclear power plant will come into being that can not only reuse nuclear waste but also enhance present electric energy production by 50 times. Second, only a small amount of nuclear fuel is necessary to generate large amounts of energy; its energy density is millions of times greater than that of fuel, and the energy released by 1 kilogram of uranium is equivalent to that released by 2400 tons of standard coal. Third, the carbon dioxide emission per kilowatt-hour of a nuclear power plant is only one-fiftieth of that by a coal-fired power plant. Finally, nuclear energy power generation is low in cost; its basic construction investment is generally 1.5–2 times that of an equivalent thermal power plant, but has much lower cost fuel. Its operation and maintenance cost is also less than that of a thermal power plant. Only 5% of electricity price per kilowatt-hour comes from uranium in nuclear energy power generation.

Safety is of vital importance in developing nuclear energy. The harm inflicted by atomic bombs and the disastrous consequence of a possible nuclear leakage makes people unwilling to use nuclear energy. After the nuclear accidents at Three Mile Island and Chernobyl, and the nuclear accident at Fukushima caused by an earthquake, new safety concerns regarding nuclear power plants are being taken into consideration more seriously; they have made every nation reexamine the development strategy of nuclear power. Facts show that people are still afraid of nuclear power. They worry about the chance of a serious Chernobyl-like nuclear accident or a terrorist attack at nuclear power plants. Besides, the problem of nuclear waste that is radioactive and harmful to living things and whose half-life period is as long as thousands of years, tens of thousands of years, and even hundreds of thousands of years still restricts nuclear energy development.

Though very controversial, accelerated development of nuclear energy has become a big trend in many countries (Figure 7.7). China plans to build over 30

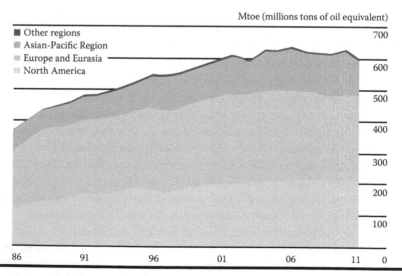

Figure 7.7 Global consumption of nuclear energy. (From BP World Energy Statistics 2012.)

nuclear power plants with a total installed capacity of 40,000 MWs by 2020; India is building 9 nuclear power plants with a total installed capacity that will increase to 20,000 MW from the present 2.5 MW; Russia similarly has spared no efforts and has restarted the frozen nuclear power plant projects after the Chernobyl accident; the United States, as the leader of nuclear energy development in the world, also passed the Act of Redeveloping Nuclear Energy in 2005 and has taken many measures to eliminate investors' misgivings. Clean solar energy, wind energy, and geothermal energy are very limited in scale of use and conventional fossil resources are depleted with each passing day, so development of nuclear power plant seems to be the first choice.

Nuclear Waste Tag—10,000 Years

A team of experts from different fields visited the Los Medanos salt mine one morning in the summer of 1991. The mine is 655 m underground and is situated in a nontraversed desert center in New Mexico, which is 42 km away from the nearest city, Carlsbad. Why were these outstanding linguists, anthropologists, archaeologists, material engineers, and even science fiction writers visiting the cave? It was because they were invited by the U.S. Department of Energy to join in a specially designed think tank.

The salt mine was in fact a "Waste Isolation Pilot Plant" (WIPP) with strategic meaning and would receive the waste left by the U.S. in making nuclear weapons. Indeed, the U.S. has accumulated thousands of machine tools and containers polluted by plutonium or americium since the Manhattan Project and the first atomic bomb in 1942. The amount was enough to release alpha rays for tens of thousands of years. The salt mine will be thoroughly closed by 2033 when the buried waste reaches 176,000 m³ (about 7% of the volume of the Great Pyramid). Experts guarantee that the isolative and protective role of the rock salt will last for at least 250,000 years.

However, the first global plan of burying radioactive material arouses people's anxiety. What should later generations do if they unintentionally open the "Pandora's box"? The region is famous for potassium mines and contains rich petroleum and natural gas resources, which will possibly attract future explorers. In short, there are many factors that will result in disastrous consequences. This forces authorities to assign a task that is slightly absurd: to warn our descendants that even after 10,000 years, it is still very dangerous to dig and drill around this area.

As a consequence, the experts mentioned earlier were invited to go into the salt mine and think of how to set up warning signs for the "dangerous articles." They had to work out information that would be readable and understandable by the people 400 generations later no matter what changes have been made in human knowledge and culture. The experts found the task very difficult.

Although most countries have begun to think about the problem of deeply buried nuclear waste, WIPP is still the only radioactive waste disposal site that has been utilized in the world so far. Even though we are extremely careful in choosing sites that are far away from resources, these sites are still faced with the same challenge: to withstand oblivion! Challenges involved might not be solved merely by burying nuclear waste in a rock cave because the poison produced in incinerators and heavy metals accumulated in factories have lasting impact. The warnings set up by the present safety organizations or enterprises will need to last for thousands of years. The experts made an assessment of the historical memories of human culture, and their conclusion was that these memories could be maintained for at most 500 years. Of course, the Vatican keeps the data of 8 centuries, but this is not worth mentioning compared with the warning sign required by WIPP, which needs to last for at least 100 centuries!

(Referred from the article with the same name in *New Discovery* magazine, February 2009).

The Challenges We Face

Science, technology and productivity have developed in leaps and bounds since the Industrial Revolution, especially since the 20th century. Development has become the loudest bugle call played by the whole of mankind. But the problems caused by unbalanced and unsustainable development have begun to appear—global warming, ecological environment pollution, fewer and fewer natural resources (especially fossil energy), and energy disputes. These predicaments, directly or indirectly related to energy, have become serious challenges that we cannot dodge.

Chapter 8

Global Warming: A Hot Topic

8.1 Sea Level Rise

An invisible blanket has formed due to the carbon dioxide emission from burning of fossil fuels such as, coal, petroleum, and natural gases—after the Industrial Revolution, and especially in recent decades—and has trapped the heat from solar radiation on earth itself (not allowing it to diffuse to outer space). This has led to the earth's surface warming—the greenhouse effect. The global climate change owing to this has brought many disasters to humankind and the ecological system, such as extreme weather, glacial ablation, permafrost melting, coral reef death, sea level rise, ecological system change, drought and flood disasters, deadly heat waves, etc.

Marigraphs all over the world show that the global sea level is continuously rising due to global warming, in some cases even threatening human survival. The sea level has risen by 20 cm since the early 20th century, and especially more in recent decades; this rate has doubled from an average annual increase of 1.8 mm in the 20th century to 3 mm at present, and is still increasing. This trend has broken the general stability* of the sea level, which has been maintained for 3000 years. What will it lead to? We do not know for certain.

The greenhouse effect results in sea level rise for the following two reasons. First, water temperature rise causes the sea's volume to increase. Most solids, gases, and liquids have the physical property of thermal expansion. A quarter of the sea level rise was caused by oceanic thermal expansion before the 1990s. Second, the melting of the glaciers on high mountains, the ice sheet on Greenland, and the

---

* See "When Sea Water Goes Up Three Meters," by We Zhenghao and published in *Scientific Showplace* in December 2009.

ice cap on the North and South Poles has resulted in sea water increase. In fact, the melting of the ice sheet on Greenland alone increases the sea level by about 0.21 mm annually. The ice caps on the North and South Poles have also started collapsing uncontrollably in recent years (see Figure 8.1). IPCC* predicted that the global sea level would rise by 20–60 cm, but they did not take the current situation of the melting ice caps into account.

As the glaciers on the North and South Poles collapse, 500 billion tons of ice flows into the sea. Its disastrous consequences lie not only in their resulting in the sea level rise but also in their increasing the frequency and strength of natural disasters. Coastal regions are affected by windstorms each year, and their frequency and strength differ at different places and based on climate. Storm surge formed under the joint action of windstorms and low pressure impels sea water to go upward, making ocean waves rise several meters higher than their normal level within several days and even several hours. Such water surges with a height of over 10 m and a weight of hundreds of tons will have a great impact on coastal regions.

In recent decades, people have started moving to coastal regions all over the world. At present, 20% of the global population lives within 30 km from the coastal line; and the number is still increasing. The phenomenon is attributed to many reasons, such as, the increase of rich retirees, fertile land of the delta regions, and

Figure 8.1   Ice cap collapses under the action of gravity.

---

* IPCC, Intergovernmental Panel on Climate Change.

beautiful scenery of coastal regions. The potential threat of windstorms is more severe owing to such a trend. Moreover, there is no evidence that the sea level rise would be limited to 60 cm as predicted by the IPCC, and so uncertainty of disasters greatly increases.

In 2007, the World Bank conducted research to determine how many properties, farmlands, and people in the 84 developing countries would be affected if the sea level rises 30 cm and 50 cm. The research shows that if the sea level rises 30 cm, 135 million people in China will need to relocate, 3.2% of GDP and 2.5% of urban areas will be lost, and 1.1% of farmland will be flooded; if the sea level rises 50 cm, 300 million people will become homeless. These data are probably greatly underestimated because the research did not take windstorm and other disasters into consideration.

Nature has repeatedly exhibited its unpredictable power. In August 2005, a category 5 hurricane—Katrina—from the Caribbean landed in America, bringing disaster to a region the size of the British territory, taking 1500 lives in New Orleans and the Gulf Coast, and making hundreds of thousands of people homeless. In May 2008, Cyclone Nargis made landfall in Myanmar at a speed of 192 km/h, killing nearly 100,000 people, and making 500,000 people homeless. In October 2012, Typhoon Son-Tinh and Hurricane Sandy wreaked havoc on the eastern and western hemispheres one after another. In the eastern hemisphere, at the South China Sea, Typhoon Son-Tinh went north–northwest all the way after it formed in the northwest Pacific in the southeastern Philippines. It gradually intensified and upgraded from a strong tropical storm to a strong typhoon as it encountered the a strong, cold northern air, resulting in heavy rain, which led to mountain torrents and serious damage. In North America, Hurricane Sandy, swept through the American eastern coast, and brought a rainstorm that flooded Atlantic City. In America's biggest city, New York, where the UN headquarters is located, 7 railways were inundated, 50 residents in Queens were caught in a fire, and there was a blackout throughout the city. The hurricane swept through 14 American states, affecting at least 60 million people, causing a loss of US$100 billion, surpassing the loss brought by Hurricane Katrina in 2005.

Similar tragedies may be more frequent and serious. "It is not a problem of whether there will be a hurricane attack, but a problem of when it will attack" (LAgence France-Presse, AFP, May 25, 2006). Max Mayfield, director of American Hurricane Center said in May 2006 when he was interviewed by a reporter of Agence France-Presse. If the disaster had occurred in densely populated cities or regions, it would have imperiled the safety of tens of millions of people and destroyed property; such a great disaster is unimaginable and unbearable.

The five countries likely to be submerged due to unprecedented sea level rise are as follows:

Tuvalu (see Figure 8.2), a very small island country in the South Pacific, consists of 9 circular coral islands. Looking down from the air, it looks like a long and narrow

Figure 8.2   Funafuti atoll beach in Tuvalu. (Photo by Stefan Lins.)

sea wall; the land area is only about 26 km². Because of the temperature and sea level rise, Tuvalu's Funafuti atoll coastline has shrunk about 1 m inward in recent years as its highest elevation is only 4.5 m above sea level. At present, the average width of the land on the main island is only 20–30 m; at its widest, it is only a few hundred meters across. It will become the first nation that sinks into the sea because of climate change, if the sea level continues to rise.

Maldives (see Figure 8.3) is an island state in the Indian Ocean and a world famous tourist attraction; it is called "paradise on earth." Maldives consists of 26 natural atolls and 1192 coral islands. When viewed from the sky, it resembles a string of pearls. Its average elevation is only 1.2 m, and most of the country (80%) is not more than 1 m above sea level. But, the paradise on Earth may be submerged by seawater in the near future. IPCC has predicted that the whole of Maldives will be completely under sea level by 2100. Mohamed Nasheed, President of Maldives, stated: "our country will become a dead area if the temperature rises by 2°C. As president, I can't accept it; as a citizen, I can't accept it even more."

Kiribati (see Figure 8.4), the only nation in the world straddling the equator and bordering the International Date Line, is situated in the middle of the Pacific and is made up of 33 islands, with an average elevation of less than 2 m each; the islands are mostly low and flat coral islands. President Arnott of Kiribati, says: "for

Figure 8.3   An aerial view of Malé, capital of Maldives. (Photo by Shahee Ilyas.)

Figure 8.4   A bird's eye view of Kiribati. (Photo by Edvac.)

many years, we have suffered from the flood brought by large sea waves and tides. In recent years, because of the serious erosion from sea, the phenomenon of a whole village moving has occurred to many regions in our nation, our crop production is often devastatingly stricken, and intruding seawater has polluted the freshwater on the islands." Kiribati will face catastrophe if the sea level rises by 40–80 cm. For this, Kiribati has prepared for the worst: All the people in the country will move to another country if necessary.

Bangladesh (see Figure 8.5) is situated on the delta area in the northeast of the South Asian subcontinent formed by the floodplains of the Ganges and Brahmaputra, close to the Bay of Bengal in the south. Most of the territory is low and flat flood plains with an average elevation <10 m above sea level. With climate change, the melting speed of the glaciers on the Himalayas accelerates. Ninety percent of the melting water will flow into the Ganges delta and bring floods to Bangladesh. Dieppe Moni, Minister of Foreign Affairs of Bangladesh, says: "If the sea level rises by 1 m, 30% of Bangladesh territory will be submerged by seawater, resulting in 40 million people homeless in the end." It is estimated that by 2050, 20 million people in Bangladesh will have been forced to leave their hometown.

Vietnam (see Figure 8.6) is located in the east of the Indo-China Peninsula, bordered by South China Sea in the east and south. The sea level rise caused by global warming will also have a serious effect on Vietnam. If the sea level rises by 1 m, three-quarters of the Red River plain and the Jiulong River plain will be inundated by sea water, and one-tenth of the total population will be affected.

Figure 8.5 Bangladesh during flood. (Photo by Rezowan.)

Figure 8.6   Vietnam during flood. (Photo by RG72.)

The Department of Natural Resources and Environment of Vietnam states: "Climate change has affected Vietnam; and floods, typhoons, drought and other natural disasters are increasingly frequent and serious. Although we have done our best to lessen all kinds of losses, hundreds of Vietnamese still die of natural disasters each year, resulting in economic loss of tens of thousands of US$. Besides, the risk will be greater and greater."

8.2 Seasonal Chaos Disrupts the Biosphere

Seasonal chaos is regarded as another adverse consequence resulting from widespread use of fossil fuels. An increasing number of studies indicate that the features of the seasons are changing—winter without snow, spring coming in advance, summer too hot to endure and so on. Besides, the changes are closely related to climate change. More importantly, seasonal chaos disrupts the biosphere, and endangered animals and plants all have to try their best to adapt themselves to new environmental changes.

Astronomically, seasonal alternation occurs with change in sunshine angle and time and depends on the motion of celestial bodies (see Figure 8.7). For example, the June solstice (June 21 or 22) is the longest day in the northern hemisphere. The sunshine and the oblique angle of the earth's rotational axis are basically stable, and so the seasonal alternation follows the prescribed order without change, no matter whether there is climate change or not. But, meteorology and climatology both do not give objective standards to define which period of the year is winter and which period of the year is spring, and a year is divided into hot and dry seasons and cold

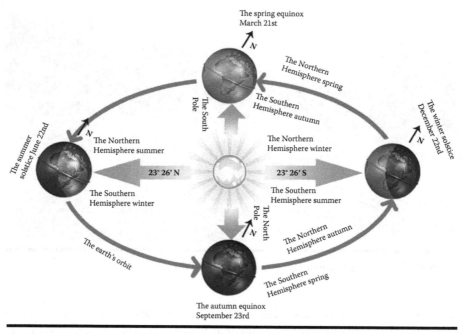

The spring equinox
March 21st

The Northern
Hemisphere spring

The Southern
Hemisphere autumn

The Northern
Hemisphere summer

The Northern
Hemisphere winter

The winter solstice
December 22nd

The summer
solstice June 22nd

The South
Pole

23° 26′ N

23° 26′ S

The Southern
Hemisphere winter

The Southern
Hemisphere summer

The North
Pole

The Northern
Hemisphere autumn

The earth's orbit

The Southern
Hemisphere spring

The autumn equinox
September 23rd

Figure 8.7   Revolution of the earth and seasonal change.

and rainy seasons from the point of view of climate; between the seasons are transitional periods. Whether the transitional period is long or short is dependent on the conditions of topography, elevation, vegetation, and even the sea in different places.

The difference of duration of main seasons and transitional seasons of different places brings many difficulties to meteorology and climatology that study weather change mainly with statistical methods. For comparison of similar data (temperature, rainfall, etc.), we need to divide a year into the four seasons with the months closest to the astronomical seasons. Computer simulation data clearly show that the "normal" average temperature (from statistical data of recent 30 years) of the different seasons will go up by 4°C–6°C by about 2100.

At present, the winter that we have long been accustomed to is gradually disappearing. The icy period of winter begins later, ends earlier and might finally disappear. Even temperature increases of 2°C (possible before 2050) will reduce snow-covered areas by 40%–50%. Snow is already rare in plain regions at present, and it will become a distant memory in the future. As winter slowly disappears, summer will come earlier. If present climate change trend continues, there will be less rain in future summers; the best scenario will be that the present level of rainfall is maintained, but it will be hotter. Computer simulation and prediction of climate indicates that the drought in future summers will be more and more severe and that temperature will rise by a big margin. Although the water stored in the soil and vegetation is of extremely high thermal capacity and plays a significant role in lowering temperature, once it completely evaporates, nothing can stop the heat waves.

Now that cold winters gradually become warm and summer is increasingly hot, spring and autumn also do not remain unchanged; they are similarly affected by climate change. With the remains of the summer drought, the "autumn tiger" can show off its power for some more days, and spring, affected by even a slight change in climate, is more difficult to distinguish from winter. Besides, as the first heat wave comes earlier each year, it is not known how much time is left for the "spring goddess."

Although the final conclusion has not yet been reached on the trend of the future change of the transitional seasons in meteorology, it is evident in biology: Spring begins earlier and earlier, but autumn begins later. We can study the change of the four seasons through the growth features of deciduous plants. Spring begins with plants germinating and ends with plants flowering, and autumn begins with the cessation of plant photosynthesis and ends with the fruit being fully mature and leaves withering and falling off.

The reason for biological changes is global warming, and this leads to changes in the duration of the four seasons and in nature. Climate change also breaks the normal growth law (see Figure 8.8) while interrupting the seasonal "clock" and changing seasonal features.

The ecological balance between different species will be broken because the potential of their adapting to new environments and diversified development is different; biodiversity will also change. At present, it is difficult to make relatively

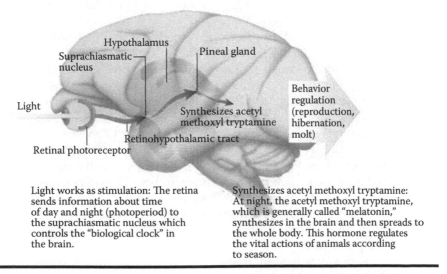

Figure 8.8   The principle of animals' "biological clock" controlled by light.

precise predictions for the future in light of the complexity of the ecological system, but many species may disappear forever. It is difficult to answer, "Does the pendulum of nature on earth need to be corrected and when it will be corrected at present?" It is still difficult to answer.

Animal Behavioral and Physiological Disorders

Animals' reaction to the seasonal chaos, especially to spring coming earlier, differs greatly. A conclusion can be reached by careful observation of animals' migration, hibernation, reproduction, and other seasonal behaviors that some creatures benefit from seasonal change due to their adaption to it; other creatures are just the opposite.

Migration. For birds, migration is first a survival tactic for dealing with the predicament of food shortage in winter while returning to their reproductive habitat, when there is abundant food, for giving birth to their young. Yet, the advent of spring in advance has struck a heavy blow to birds. Take *Ficedula hypoleuca* (see Figure 8.9) as an example. The birds did not change their migration time like other migrating birds. As a result, when they return to their wintering ground in southern Saharan Africa, spring had come earlier. Even though adult birds laid eggs sooner, newborns still did not have enough food to eat.

Hibernation. The research carried out by the Rocky Mountain Biological Laboratory in the United States during 1975–1998 shows that as spring starts earlier, precisely ending hibernation has had a vital effect on animal survival. *Marmota flaviventris* (see Figure 8.10) generally wakes up at the end of winter, digging a tunnel in the snow to "detect" outside temperature to decide when it will end its winter sleep. In 1998, it ended winter hibernation 38 days earlier than 23 years ago. This was related to the temperature rise of 1.4°C in the spring of that year compared with the previous one. The temperature rise also resulted in more snow, so the snow was thicker when the *Marmota flaviventris* woke up. It had to risk the danger of dying from hunger, waiting for a longer time before it could find green grass.

Reproduction. Salmon (see Figure 8.11) in the Atlantic go to the Neville River in the Basque region for laying eggs between

Figure 8.9 *Ficedula hypoleuca* **cannot find enough worms to support itself and its young because of seasonal chaos. (Photo by Gegik.)**

Figure 8.10 *Marmota flaviventris* **face the danger of dying from hunger. (Photo by Nick Comer.)**

Figure 8.11 **The Atlantic salmon find it difficult to lay eggs because of warm winters.**

January and February each year. The research of the French Academy of Sciences indicates that warm winter is a threat to them. The water temperature needed for the salmon to lay eggs is generally not above than 12°C, but the water temperature in the Neville River in winter is, currently, frequently above 12°C. Salmon in the Neville River might become extinct if the water temperature is above 12°C for three consecutive years.

Chapter 9

Environmental Deterioration: Not an Unwarranted Threat

9.1 Troubling Garbage

The environment we rely on for survival has deteriorated drastically, and the fundamental reason is the development and use of energy; it can even be said that the core of the environmental problem is the energy problem. The development and use of energy usually cause damage to the environment, leading to numerous associated effects; so, upon final analysis, all environmental problems are closely linked with energy.

The rapid development characterized by industrialization and urbanization leads to the production of a large amount of household garbage and industrial waste. The garbage problem has spread from cities to the countryside and brought serious problems to human life and survival.

Pollution caused by household garbage is mainly divided into air pollution, water pollution, and biological pollution. Household garbage piled up in the open air produces large amounts of ammonia, sulfide, and other organic matters. Volatile, soluble, and harmful gases cause air pollution. Household garbage also contains pathogenic microorganisms, produces large amounts of acids and alkaline organic pollutants in the course of piling due to putrefaction, and release of heavy metals. Hence, garbage forms a source of pollution containing organic matter, heavy metal, and pathogenic microorganisms; it pollutes surface water and underground water if it is not disposed in time. Besides, garbage dumps are the breeding place of mosquitoes, flies, cockroaches, and rats; thus, helping spread diseases and being a threat to human health.

Industrial waste contains many kinds of poisonous and harmful matter. Among them, organic pollutants include chlorinated hydrocarbons, hydrocarbons, etc.; inorganic pollutants include mercury, cadmium, lead, arsenic, zinc, chromium, etc.; and physical pollutants include radioactive pollutants, etc. These pollute the soil, air, and water and harm human health through many channels.

Currently, electronic garbage has become a severe global environmental problem. It mainly contains lead, cadmium, mercury, and other elements that are harmful to our bodies. Take computers as an example; it takes over 700 kinds of chemical compounds to make a computer. Among them, more than 300 materials are harmful to humans. A computer's display unit contains lead, which is harmful to nervous system, circulatory system, excretory system; its battery, case, and disc drive contain chromium, mercury, and other elements. Chromium compounds can cause asthma, and mercury is destructive to human cell DNA and brain tissues. The harmful materials contained in electronic garbage will seep underground and lead to underground water pollution if they are buried randomly; they will release poisonous gases and pollute the air if they are burned. What is more serious is that the heavy metal contained in electronic garbage will bioconcentrate,* resulting in lasting pollution through food chain transfer, which is permanent once the pollution has begun.

Construction waste accounts for a high ratio of municipal waste in developing countries. Chinese construction wastes, for example, accounts for 30%–40% of municipal waste. The floor area in China will increase by 30 billion m^2 by 2020. If $10,000\,m^2$ produces 500–600 tons of waste, the amount of new construction waste will reach shocking proportions. At present, most construction waste is generally transported to the outskirts of town or to the countryside for piling or landfill without strict, safe disposal. The adhesive, coating, and paint in construction waste contain high-molecular-weight polymers that are not easily biodegradable; they also contain heavy metals. These can cause underground water pollution and directly harm surrounding residents if they are buried underground.

No doubt, it is correct that "waste is the resource misplaced." Old furniture, newspapers, magazines, cans, plastic bottles, beer bottles, and other waste that are easily recycled have been recycled since early times by junkmen (see Figure 9.1). But, the question is not how to convert the waste into resources but how to dispose of it.

Landfill is still the main method used globally to dispose of waste. China has adopted a system of hygienic landfill. It lays stress on isolation from the environment, in the sense that the bottom of the landfill is isolated so as to not cause pollution of underground water and the top is covered so as to not release odor.

---

* Bioconcentration is the phenomenon by which biological organisms or the biological population accumulates certain elements or irresolvable compounds from the surrounding environment and increases the concentration of the material to a level higher than that seen in the environment.

Figure 9.1   Garbage classification in some rural areas.

But this raises a new question: the waste is now in anaerobic conditions in an isolated space—better isolation means slower degradation speed and longer periods of potential harm to the environment. This affects land use after the landfill is closed. Thus, there exists a real paradox in hygienic use: The release of pollutants in the landfill will perhaps take over 100 years, but the reliable life of the engineering materials used cannot last that long. Moreover, people have the dilemma of how to assess and make a choice regarding the engineering materials used for isolation and the release period of the pollutants into landfills.

The History of Garbage Disposal

Since ancient times, almost every society has faced or endured the problem of waste disposal. Solid waste, which is the most common, generally accumulates in large quantities and is the most difficult kind of waste to dispose. In ancient Troy, waste was sometimes discarded indoors or dumped on streets. When the smell at home became unbearable, some earth would be fetched to cover the waste; or the pigs, dogs, birds, or rodents were allowed to eat the waste. According to recorded data, the waste of Troy accumulated to up to 1.5 m in height every 100 years, and even reached 4 m in some places.

Garbage chutes and garbage cans began to be established in the houses in Mohenjo-daro in the Indus Valley, according to a uniform plan, in about 2500 BCE. The waste in the aristocratic district of Heracleopolis, Egypt, began to be collected and dumped into the Nile in about 2100 BCE. The bathrooms of some houses in Crete, Greece, had been connected with main sewer lines at about the same period. Special land was allocated by the island for waste disposal in 1500 BCE.

Jews began to bury waste at places away from residential areas in about 1600 BCE. Religion also played a certain role in the carrying out of hygienic measures. The *Talmud* prescribed that the streets in Jerusalem had to be washed, although the water available in Jerusalem was very limited.

The suburbs in Athens were full of garbage in the 5th century BCE, threatening the health of the citizens. So, Greeks began to set up urban garbage landfills. The Athens parliament decreed that dustmen had to dump waste at a place not less than 1.6 km away from the city wall. It also promulgated a decree to ban people from discarding waste on streets (the first known decree of this kind). Athenians still set up composting pits. Ancient Maras in Peru discarded waste in landfills.

The ancient city of Rome faced an unprecedented hygienic problem, which had not been seen in ancient Greece those days, because it was very large and densely populated. Rome had a strong sewage disposal system, but its disposal of solid waste was not well established. Although waste collection and disposal were organized in accordance with the standards then, the demand still could not be met. The city's waste was collected only when the nation held special activities. Landowners were responsible for sweeping adjacent streets. But these provisions were not fully carried out.

At the end of the 12th century, large numbers of people surged to cities. Streets began to be built and swept in these cities. Streets began to be built in 1184 in Paris, but the cost of sweeping the streets was not paid for by the public until 1609. Pigs, geese, ducks, and other animals were also transported to cities with people. In 1131, Paris passed an edict, prohibiting pigs from running about on the streets. It was formulated because young King Philip died in an accident caused by an untended pig.

The sanitation system in large Muslim cities and China was the most advanced in the world until the late 19th century. The European sanitation system was not developed in ancient times; it slowly improved through the Middle Ages and the Renaissance period, but obviously worsened with the rise of the Industrial

Revolution in the 18th century; Lewis Mumford* described it as "the worst urban environment in the world so far."

More and more people moved to industrial centers, which could not provide housing for them; serious crowding and health problems arose. In 1843, 212 people used one toilet on average in one district (Manchester) in Britain. Thus, a sure link between disease transmission and dirty environment was found. Massive municipal infrastructure and public health organizations began to be set up to tackle the urgent health problem.[†]

9.2 Formidable Crust

It is unimaginable that humans can affect the earth and cause earthquakes, is it not? Can earthquake be attributed to humankind? How can evidence be found? The studies of decades make the scientists who are in charge of monitoring earth pulse worried. Their worry originated from the real concurrence between human behaviors, such as deep well drilling and water storage in reservoirs, and earthquakes both in time and in space. We would doubt the reality of the connection if we had observed only an earthquake. But a series of earthquakes took place at specific times and places, and so we have to consider that we may have influenced them.

To verify the influence of human activities on geological structure, we must precisely understand the historical earthquakes of certain areas. France has maintained a good record of its earthquakes with tremors over the past 500 years. Earthquakes, including earthquakes above magnitude 4, observed at the Pau region, France, since the 1980s were all caused by massive exploitation of natural gas. The record shows that the earthquakes first appeared 10 years after the first exploitation of natural gas, and frequently occurred from then on. Northeastern America was originally a seismically nonactive region, but a series of earthquakes have taken place there since the 1980s. The common view obtained from studies is that one-third of these earthquakes were directly caused by exploitation of deep mineral resources, large-scale quarries, and hydraulic wells.

If an area is not originally an earthquake-prone area, it is easier to find the link between the earthquake and oil exploitation or storing water in a reservoir. The same goes for detecting earthquakes caused by underground nuclear experiments, such as the nuclear experiment performed by North Korea in October 2006, which led to a magnitude 4.2 earthquake. Compared to earthquake-prone areas, earthquakes taking place in the "silent" continental area are more greatly influenced by

---

* See *Waste* by M. V. Melosi and translated by Zheng Qiaoshan. Foreign Language Teaching and Research Press. November, 2004

† Lewis Mumford (October 19, 1895 – January 26, 1990) was an American historian, sociologist, philosopher of technology, and literary critic.

human activities because they usually occur on the top of the earth's crust, which is within the interrupting scope of human activities. The ecological balance here can more easily be damaged by human activities. The Mohr–Coulomb rock mechanics theory, with a history of over 100 years, can help describe how a fault takes place due to stress change. The vertical force produced by the mass of the top rock of a fault and the horizontal force produced by plate motion are two kinds of mutually opposite stresses. The rock deforms under the action of these stresses and fractures when the stresses exceed the endurance of the fault and releases its stored energy—this is an earthquake (see Figure 9.2).

Humans build large-scale reservoirs to store huge amounts of water or drill holes to extract millions of tons of minerals or hydrocarbons, and these are heavy loads added to or reduced from the crust. It changes the stress on the fault; this, along the action of the force produced by plate motion, provides the right condition for fault fracture.

In other words, the earthquake is a kind of natural phenomenon, and human activities play an inducing role, causing the fault fracture. Constructions near fracture zones are very likely to activate fault zones, thereby causing earthquakes.

If human activity can lead to fault activation, this means that humans also have the ability to contend with the power of the earth itself, which is quite possible. We must recognize that earthquakes can be induced by human activities; no longer can the earthquake be regarded as a completely inescapable natural disaster from which we can shirk our responsibility. What we need to study further is to what extent humans accelerate fault fracture. Human interruption only makes the earthquake take place several years ahead of time than if the fault was originally approaching the fracturing area at a high speed, such as, at the edge of a geological plate e.g., Pacific ring of fire or Himalayas). In areas, such as the middle zone of a geological plate (e.g., South Africa, Australia, or North Europe) where the fault is in extremely slow motion, it would take thousands and even tens of thousands of years for an earthquake to occur. In such case, it can be basically inferred that the earthquake would not take place at all if there was no human interference.

In such a troubling situation, someone has proposed a question the answer to which could help us take preventive measures: Now that it is known that human activities can induce an earthquake, can we predict the date on which an earthquake takes place? Unfortunately, the answer is negative. An earthquake, whether natural or artificial, cannot be predicted. The reason is simple: The action of the stresses in the crust cannot be identified through observation.

Human life and development increase the demand of energy and other resources with each passing day; so reservoirs are built larger and larger, well-drilling ability becomes more and more powerful, and the influence of human activities on nature becomes more and more proactive. Although it cannot be asserted that human activities at fault zones will lead to an earthquake and it is impossible to completely stop related activities, it is necessary to recognize the weakness of the crust and proactively restrict and reduce human activities at geological fracture zones.

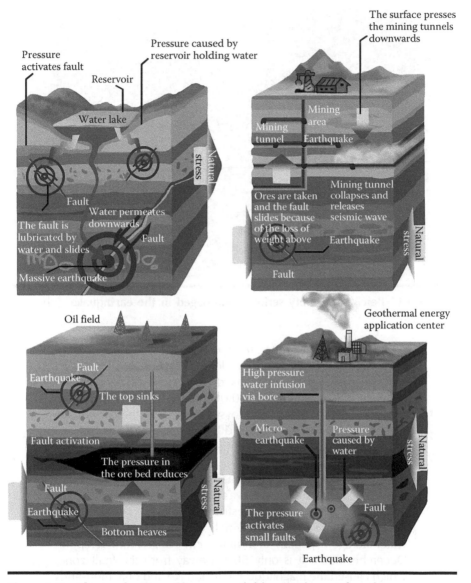

Figure 9.2   The mechanism of human activities causing earthquakes.

The Violent Earthquake in Wenchuan, China

The great earthquake (see Figure 9.3) that occurred in Wenchuan, China on May 12, 2008, was one of the most devastating earthquakes in recent decades. Mountains and rivers

Figure 9.3 Beichuan County seriously damaged in the earthquake. (Photo by Fan Xiao.)

were displaced, the ground trembled, people were parted from each other by distance or by death; a scene of devastation met the eye everywhere; tens of thousands of people died or were missing, nearly 400,000 people were wounded, 5 million houses ruined. The losses caused by the earthquake were equivalent to those caused by the Indian Ocean tsunami that occurred in December 2004 and Hurricane Katrina that occurred in August 2005, listed as the most horrible natural disaster in history. After the disaster, some people thought that the Wenchuan Earthquake was perhaps a sign that human activities are of potential destructive power. The Zipingpu Reservoir (see Figure 9.4), which is 156 m high and has a water storage capability of 1 billion cm^3, began functioning in December 2004. It is only 500 m away from the fault zone, which caused the magnitude 8 earthquake and only thousands of meters away from the epicenter. It happens to be situated at the zone where the Indian plate impacts downward toward the Asian–European plate and the zone where the Tibetan Plateau and Sichuan Basin meet.

There are also lots of people who believe the Wenchuan Earthquake was just the inevitable result of geological plate motion. There is no final conclusion for the dispute because it cannot be verified by repetition. That human activities induce earthquakes is a fact that is generally accepted, although its

Figure 9.4   Zipingpu Dam.

probability is very low. Therefore, it is necessary to reexamine and scientifically reassess important engineering constructions, such as, drilling, mining, and storing water.

People should calm down and think carefully whether the natural disasters that have come one after another recently are accidental or geological responses to human activities.

9.3 Endangered Rivers

Water is the source of life; rivers are the mother of civilization. Water is the basic matter that makes up organisms and the main medium of metabolism; its demand is very high for industrial and agricultural production. It is also a very important environmental element that can alter climate and promote landform formation. Water as an irreplaceable resource, dominates humankind's economic and social development and the progress of civilization.

Yet, the water resources that can be used, especially fresh water resources, are not what we would call "inexhaustible." The oceans occupy an area of 361 million km², while the land occupies an area of 149 million km² on the earth; three-quarters of the earth is covered with water; so it is called a "watery planet." There is 1.386 billion km³ of water altogether on the earth. At first sight, this seems like a big number, but after a careful analysis, the sea water that cannot be directly be used for our life and production is 1.338 billion km³, accounting for 96.5%, land water of 48 million km³, accounts for 3.5%, of which fresh water is 35 million km³, accounting for 2.5% of the all the water on the earth. About 68.7% of fresh water is distributed in the glaciers of the North and South Poles, permanent glaciers, and snow on the high mountains of the various continents; only 31.3% of the fresh water that can be used for human life and development is distributed in rivers, lakes, reservoirs, soil, and underground aquifers, accounting for 0.78% of all global water, which is a very small quantity indeed.

Rivers are the main carriers of fresh water and they nurture life and civilization. The civilizations of ancient Egypt, Mesopotamia, India, and China originated from aggraded valley plains at the banks of rivers that are called "Mother River" by local people. Vast aggraded valley plains and endless river fresh water resources formed ideal conditions for the massive survival and development of human society. Human activities are closely linked with the rivers, and the rivers greatly affect the survival and development of human society. It is imperative that to benefit people now and in the future, we must abide by objective laws, transform, develop and use rivers sustainably; otherwise, we will feel the destructive power of the rivers unleashed.

Since early times, every country has taken great efforts in transforming and developing rivers, using river basin land, and utilizing river water for irrigation, shipping, electric generation and drinking, etc., to resolve the energy problem faced in survival and development. Therefore, the original appearance of rivers have been changed, their natural roles more and more seriously interrupted. Reclaiming floodplains at the banks of rivers has cut the links between the rivers and the land at their banks; diverting water away from the rivers has reduced their runoffs; building dams or intercepting river water has stopped or changed its flow; constructing reservoirs that have adjusted runoffs has changed the hydrological situation of the rivers; discharging waste water into the rivers has polluted river water and underground water; excessive felling of trees and pasturing has led to rapid reduction of forests and pastures, which has increased soil and water loss; discharging greenhouse gases has resulted in climate change, which has had an adverse influence on the ecological system of rivers.

The British Scientific Development asserted in 2005 that dam-constructing activities affected the flow of rivers. A river is a giant system that cannot be easily interrupted. But if human transformation and interference exceeds its self-adjusting and self-repairing ability, its natural functions will also, gradually yet irreversibly, deteriorate—drying up will be accelerated, leading to water resource reduction.

Phenomena, such as land cracks, withering of crops, lack of water for people and livestock, frequent fires, all caused by rivers drying up can be seen everywhere in all parts of the world. Half of the rivers that do not dry up are polluted, except for the Amazon and Congo. At the same time, the development of global industries, agricultural production, and growth of human population have increased water consumption, seriously aggravating the water resources tension—human water consumption increased fivefold in the 20th century. The serious shortage of water resources caused by river deterioration and increased human demand has resulted in the fact that there is a short supply of water in 60% of the places in the world; 1.2 billion people are short of water, 3–4 million people die from water-related diseases, and many countries face water scarcity. The water resource crisis will have spread to 48 countries by 2025, making 3.5 billion people beset with water scarcity.

Disputes about water have occurred between some countries and regions, such as, the water dispute in the Middle East, the Nile dispute in northern Africa, and the Ganges River and Indus River dispute among India, Bangladesh, and Pakistan. It is predicted that disputes over water resources will become one of the main triggers for massive wars of the 21st century. The trend of acceleration of rivers drying up, serious falling of the water resource quality, rapid deterioration of the ecological system, increasing shortage of fresh water resources all form a severe threat and challenge to humankind's survival and development.[*]

The Rivers About to Dry Up[†]

Storage and excessive use of water have also become the most common factors disturbing the ecological environment of rivers, while climate warming and environmental pollution have become serious problems. In 2012, the U.S. *National Geographic* magazine listed the seven rivers in the world that were about to dry up because of overuse. They included the Colorado, Murray, Indus, Teesta, Rio Grande, Syr Darya, and the Amu Darya Rivers. The reduction of these rivers has made many fresh water creatures in these rivers nearly extinct; water resources have become seriously short, and humans face a serious fresh water crisis.

The Colorado River (see Figure 9.5), situated in the southwest of America and the northwest of Mexico, provides water for 30 million people through many dams along a stretch of 1450 miles (2334 km). At present, it seldom reaches its delta and the California Gulf because its water is used for agricultural and industrial production and urban life; only one-tenth of its water flows to Mexico.

The Murray River (see Figure 9.6), as the longest and most important river in Australia, extends 1476 miles (about 2375 km) from the Australian Alps, across inland plains, to the Indian Ocean near Adelaide City. It is regarded as a "rice bowl"

---

[*] See *Leading to Higher Civilization* by Liu Jianpin, People's Publishing House, 2008.

[†] See *Global Seven Rivers about to Dry up due to Overuse*, telecast on Yahu Natural Channel, 15 January, 2012.

Figure 9.5   The Colorado River. (Photo by Paul Hermans.)

Figure 9.6   The Murray River. (Photo by Jjron.)

because its basin is the most productive agricultural area in Australia, thanks to the irrigation from its water. The river is also the main drinking water source of Adelaide and many small cities. Its retreat has led to a rise in water salinity, threatening the agricultural production along its banks. In the early 21st century, because silt resulted in closed estuaries, water disruption and diversion, and serious water flow reduction, the river was kept open by continuous dredging to guarantee its flow into the sea and lagoon near Coorong National Park. The Murray River still faces serious environmental threat, including runoff pollution. Its drying up has also affected the Darling River that flows into the Murray River, the main channel in inland areas; it has nearly dried up because of the digging and droughts in recent years.

The Indus River (see Figure 9.7), originating from India, is the main source of freshwater in Pakistan. Its water is used

Figure 9.7   The Indus River. (Photo by KennyOMG.)

for human consumption and industrial production, but more for supporting 90% of the agricultural production in this arid country. At present, the Indus River, although one of the longest rivers in the world, no longer flows into the ocean outside Karachi Port; instead, it is drying up. The water supply interruption makes Karachi city suffer from water scarcity, and many people condemn the upstream residents for their excessive use of water. It is predicted that the population of Pakistan will increase from the present 170 million to 220 million within 10 years. But as global warming worsens further, the Indus River will retreat further, and the situation of the country's future water resources will be more severe.

The Teesta River with a length of 196 miles (315 km) originates from the Himalayas and flows through Sikkim, India, and finally into the Bangladeshi Brahmaputra River. The Teesta River is called the life line of Sikkim, but it has dried up to a certain extent because it has been used for irrigation and other purposes in recent years (see Figure 9.8). Fishermen cannot make a living along its banks, and thousands of peasants have lost their water supply. Even so, India plans to build a series of dams along the Teesta River for power generation. The weight of the sediments could cause earthquakes in seismically active areas. It is said: "Reasonable distribution of the water of the Teesta River is the only approach to improve the ecological state of these regions, but this needs to be settled by negotiations between Bangladeshi and Indian governments."

Figure 9.8   The Teesta River. (Photo by PP Yoonus.)

The Rio Grande River (see Figure 9.9), one of the biggest rivers in North America, 1885 miles long (3034 km), flows into the Gulf of Mexico from southwest Colorado, through Texas, and along most parts of the U.S.–Mexico border. It is about to dry up because of massive use of water on both sides of the border. Only less than one-fifth of the water now flows into the Gulf of Mexico. In the first several years of the 21st century, the river failed to fully reach the coast, and this led to the formation of a dirty beach, which is now fenced by orange nylon fence. Algae paint the junction of the Rio Grande River and the Carlos Arroyo River green. The water level tends to be lower than the inlet pipe of Mexico City where the Rio Grande River reaches Matamoros. The farmers in Texas allege that they lose 400 million dollars on irrigation each year. The wetland of the region was an important intermediate habitat for migrant birds, but the previous beautiful scenery has gone now, making these problems more serious.

The Syr Darya (see Figure 9.10) originates from the Tianshan Mountains of Kyrgyzstan and Uzbekistan, is 1374 miles long

Figure 9.9   The Rio Grande River. (Photo by Alan Gross.)

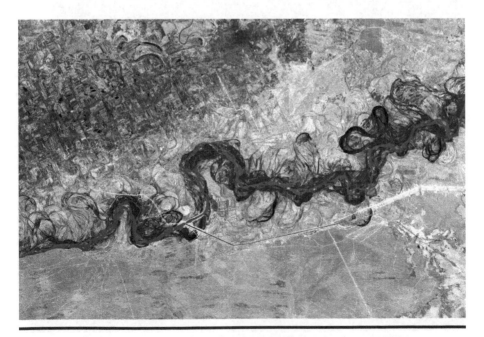

Figure 9.10   The Syr Darya. (Photo by ISS Expedition 25 crew, NASA.)

(about 2211 km), and flows into the Aral Sea. A shipping system began to be built on the Syr Darya in the 18th century and was expanded by the Soviet Union Normal University of Engineering in the 20th century. The water is mainly used for planting cotton. In fact, the river is now diverted; there are only trickles flowing into the inland sea. A dam was constructed with the financial support of the World Bank in the past several years for improving the water quality of the Syr Darya and increasing water flow to the north of the Aral Sea.

The Amu Darya River (see Figure 9.11). Lots of people have heard of the sad story of the Aral Sea. It was the fourth largest lake in the world, with a surface area of 26,000 $mile^2$ (about 67,340 km^2), making the surrounding towns flourish, supporting a lucrative muskrat fur industry and fishing industry, providing 40,000 jobs, and contributing one-sixth of the fish catches to Soviet Union. The Aral Sea originally stemmed from two rivers in Middle Asia: the Amu Darya and Syr Darya. The former was the longest river in the region and wound through the grassland for 1500 miles (about 2414 km). But in the 1960s, the Soviet

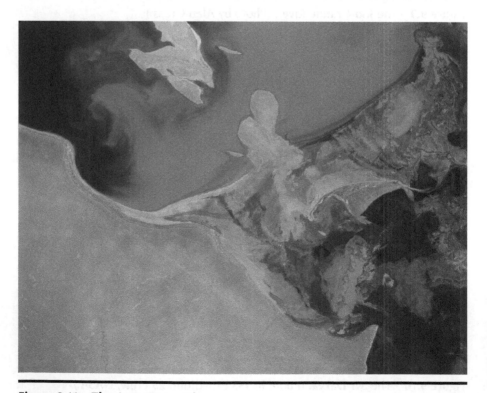

Figure 9.11   The Amu Darya River.

Union decided to develop the grasslands and set up a giant irrigation network, including 20,000 miles of canals (about 32,186 km), 45 dams, and over 80 reservoirs for irrigating the vast cotton and wheat fields of Kazakhstan and Uzbekistan. Several decades later, the water flow of the Amu Darya River has reduced sharply; only 70 miles (113 km) of its basin is left.

Lacking its main source of water, the Aral Sea shrunk rapidly. It has reduced into several small lakes in a short period of a few decades; it is now one-tenth the original total volume. The high salinity of the seawater as a result of vaporization has led to the death of millions of fish and has even caused sandstorms in surrounding areas.

Chapter 10

How Long Can Resources in Short Supply Last?

10.1 Soon-Depleted Petroleum Resources

Petroleum has been vital to the economy of all nations in the world and has been praised as "the lifeline of the modern economy." As the core fuel of global transportation networks, it provides power for billions of cars, trucks, ships, oil tankers, airplanes, and trains and supports the traffic operation of human society. However, production lags behind demand, resulting in the rapid rise of its price in the recent decade. Petroleum can never be as sufficient and cheap as before for many reasons.

M. King Hubbert,* a geologist, first put forward the concept of "peak oil" in the 1950s: The production of an oil field reaches a peak when its exploitation exceeds half of its reserves; though petroleum can still be exploited, annual oil production will be in stable but constant decline. The existing global petroleum resources can only meet the demand of about 40–60 years at the present rate of consumption. But oil consumption is still progressively increasing at 2% on average each year. At this rate, the existing petroleum resources will be depleted in about 25 years. The warning sign is when global petroleum production reaches its peak. This peak will last for a certain amount of time or will drop. But for now, as

---

* M. King Hubbert, an American petroleum geologist, first proposed the "clock-shaped curve" of mineral resources in 1949, which forms the core of the petroleum peak theory.

long as demand keeps increasing, petroleum will be in short supply, and its price will keep rising.

The naysayers generally argue that technology will save us—petroleum will be obtained from the existing oil fields and enough new oil fields will be found. But the reality is that oil production in America—technologically, the most advanced country in the world—has been falling since the 1970s. If technology really could resolve it, the problem of short oil supply should have been settled by now. Some people believe that oil production does not increase because of the restriction on oil drilling. Of course, the exploitation of some regions where oil exploitation is banned at present can really increase oil production, but the development of oil wells here is slow and results in less production. For example, the development of oil reserves in the Arctic countries needs 5–10 years, and their petroleum production might be less than 100 barrels/day, which can only meet a small part of the new demand of oil at most and not to mention to meet the demand gap due to the old oil reserves' running out. Besides, offshore areas have become an international political hotspot, which means that the development of these require longer time, higher cost, and more uncertain factors with an uncertain future. Worryingly, no matter how it is defined, the petroleum peak is rapidly coming; it has already occurred in some places. What is worse, a series of other factors will lead to oil reduction, and even a stoppage of oil production. We have consumed 2200 billion of barrels of oil in the past 100 years, an amount that had been buried in the earth's crust since the time our ancestors just began to walk upright. Light sweet crude oil that is easy to mine and refine is our first choice of fossil fuel. However, most high-quality oil has been mined out, and future oil quality will only fall. Whether we are willing or not, we can only rely even more on heavy crude oil with high sulfur content and high cost of refining, and extract it from irregular materials, such as asphaltic sands, with more money and resources. The earth has been excavated so thoroughly that little land is left to be mined.

At present, the tentacles of oil exploration have expanded into the deep sea, and so development costs have significantly risen. For example, the new "Jack 2" oil field in the Gulf of Mexico is 2000 m below sea level; oil wells are drilled 6000 m below the sea bed; a pipe of 8000 m length is needed to touch the oil layer, so the cost is very high. Carioca Oil Field in the coastal waters of Brazil is 2000 m below sea level, 6700 m below the sea bed. The challenge of building oil wells here is huge, and the cost of refining is possibly high enough to make people think twice. But, the total oil reserve of the two oil fields can only meet the world's oil consumption for a year and a half.

At present, opinions vary about when oil production will reach its peak in the world, and no unanimous conclusion can be drawn. Some people believe that with the discovery of new oil fields and progress of technology, the end will be put off to at least 2040. The relatively pessimistic people consider that the oil peak has already come silently. The International Energy Agency and the U.S. Department of Energy deem that the world oil demand will be met by the Middle East, which

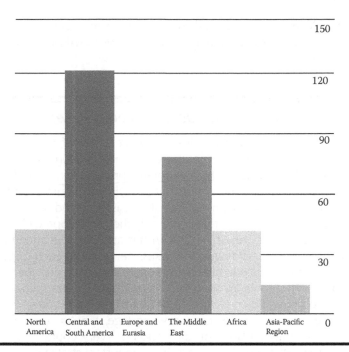

					150
					120
					90
					60
					30

North   Central and   Europe and   The Middle   Africa   Asia-Pacific   0
America   South America   Eurasia   East   Region

Figure 10.1   Global subregional petroleum reserve and production ratio.

has the highest oil production, especially Saudi Arabia (see Figure 10.1).* But if it has idle production capacity, why does it not increase petroleum production? Its petroleum production was nearly 9.8 million barrels/day at the end of 2004 when the petroleum price was 50 USD per barrel. But its petroleum production fell by 0.6 million barrels after that when the petroleum price nearly doubled. If the petroleum production fall was on purpose, it would have certainly led to the dissatisfaction of the oil importing countries, including the U.S and China, a situation no country would have welcomed. The facts indicate that the Middle East production was indeed saturated.

By the end of 2011, the proven reserves in the world totaled about 1652.6 billion barrels, which can only meet the global production demand for 54.2 years. An even worse aftermath—absolute petroleum peak—will come if no action is taken. By then, it would be meaningless to continue to drill wells and mine petroleum because the consumption for production would be more than the gain. In other words, to mine a barrel of petroleum would require the energy consumption of more than a barrel of oil to extract, refine, and transport to the filling station.

---

* Petroleum reserve and production ratio: the petroleum reserve left at the end of the year is divided by the annual petroleum production of the year, and this result indicates the number of years that the reserve can be mined if the production continues to remain.

Absolute petroleum production peak will obviously not come overnight, but if we do not take action at once, our economy, and even civilization, will likely be stopped abruptly.

The Arctic—The Last Petroleum Reserves

There are only five countries along the coast of the Arctic Ocean in the Arctic: Russia, the U.S., Canada, Denmark, and Norway. But because of high petroleum price, increasing petroleum demand, and the approaching world petroleum production peak, coupled with the fact that it is expected to be the last petroleum reserves that could help the world cope with the petroleum crisis, the Arctic Ocean has attracted the attention of the world. According to the assessment data of each of the five countries, the potential petroleum reserves can provide the world with petroleum supply for 4 months ~3 years going by the present world consumption of 86 million barrels/day.

Because of the global energy shortage, the Arctic regions have become the bone of contention of every country along its coast—its mineral resources have important strategic value, and the petroleum resources under ice are regarded as the last "feast" on earth. Not every country will have the ability to create enough alternative energy in the future years. Hence, great hope has been placed on the Arctic regions, whose extremely harsh natural conditions fail to stop petroleum companies coming to its shores one by one.

Yet, the truth is that the extremely harsh conditions make exploitation very difficult. Twenty percent of the areas that are thought to potentially contain petroleum and natural gas are covered with ice all the year round, so mining is very hard. The Arctic Ocean is fathomless, covered with ice and snow. The area where ice and snow has melted and that is open to navigation all the year around only accounts for 20% of the land; the rest is partly open to navigation in a year. However, the melting speed of the ice floes in summer has accelerated to 100,000 km^2/10 years since 1979 because of global warming. The northeast channel (Russian coast) and northwest channel (Canadian coast) that were covered with ice and mist in the past are for the first time open to navigation after 120,000 years due to accelerated melting of the ice in the Arctic.

Extremely cold climate is also a great impediment of prospecting for energy in the Arctic regions. The biggest problem caused by it is how to draw liquid petroleum to the earth's surface and prevent it from freezing—for this rigs, oil ships, storing ground, and other equipment have to be thermally insulated. Besides, land support facilities are needed for packaging and transporting the extracted hydrocarbon energy. But there are hardly any supporting facilities along the coast of the Arctic Ocean—there is neither port facility for fuel filling nor emergency rescue centers.

It is predicted that the cost of prospecting and mining Arctic energy is extremely expensive. Building a pipeline of 1000 m is as expensive as building an expressway of 1000 m; the cost of building an offshore rig far surpasses that of building the Charles de Gaulle Airport, and the expenditure of setting up and developing an offshore oil field is as high as 15 billion USD, almost equivalent to the total cost of digging the English Channel. But in accordance with the assessment of the International Energy Agency, the cost of each barrel of petroleum here will be 60 USD, its sale price will be 100 USD, so prospecting and development in the Arctic will still be profitable. Of course, all this must be based on the premise of the progress of mining technology. So far, there is only one offshore natural gas project in the Arctic that has been put into production: the Norwegian Snohvit Gas Field in the Barents Sea. In the autumn of 2007, the executor of the project, Norway State Petroleum Corporation, became the first enterprise to provide liquefied natural gas from an offshore oil field in the Arctic. The treasure of the Arctic will still be buried under its sea bed for several more years going by the present speed of development.[*]

10.2 Undesirable Coal Resources

In the last 30 years, the world coal consumption has increased about 3%/year, the same as the increase of the total energy consumption. But behind the relatively low increase of growth rate is the absolute increase in amount from 2 to 5 billion tons. It is estimated that coal alone as the fuel of thermal power generation can meet one-quarter of the primary energy demand (petroleum can meet one-third) and two-fifth of the electric demand.

The development of the coal energy industry is the most rapid among all kinds of primary energy resources. In China, called the "world factory," national

---

[*] Adapted in accordance with "The Arctic—The Last Petroleum Reserves," translated and edited by Zhang Yang and published in the first issue of 2009 of *New Discovery*.

production and life almost completely rely on coal. Seventy percent of the national power depends on coal. India is accelerating construction of thermal power stations and has made a huge plan of building the biggest coal power station in the world as a solution to frequent severe blackouts. A huge power station with power capacity as high as 12,000 MW and investment of about 8.5 billion USD will be built in eastern India. Half of America's power comes from coal power generation. There are more than 100,000 coal miners working in 2000 coal mines, mining 1 billion tons of coal each year.

Japan and Australia are also big coal consumers. Europe is unwilling to lag behind. Germany is the 10th biggest coal consumer in the world, and Belarus and Poland are, respectively, the sixth and seventh biggest coal consumers in the world. A trend of returning to coal power has swept Europe in the last 10 years because of the sharp rise of the prices of petroleum and natural gas. European thermal power stations built soon after the Second World War are being rebuilt to generate a total power of 200,000 MW before 2020. So far, most of these thermal power stations have adopted natural gas as fuel. But power producers hope to increase the proportion of coal to make the fuel structure more balanced.

In accordance with the present rate of coal consumption, existing global coal reserves can meet the demand for 150–170 years; with coal from unproven coal reserves and coal mines that are now regarded to have no mining value, we can meet the demand for 200 years, which helps ease the energy crisis greatly. Besides, the coal reserves in the main energy-consuming regions are roughly the same and guarantee status quo in geopolitics: the American continent, Eurasia, Asia-Pacific region account for 30%, only Middle East accounts for relatively less coal resources. The United States the biggest economy, has one-quarter of the global coal reserves, and China and India, the two newly emerging economies, also have very rich coal resources.

However, every coin has two sides. The carbon dioxide released by coal burning is 35% higher than that released by burning petroleum and 72% higher than that by burning natural gas. Besides, the derivatives of coal liquefaction also emit extremely high carbon dioxide. There have been talks that the world carbon dioxide emission be reduced to one-quarter of the present by 2050 to avoid the occurrence of climate disaster. But, the carbon dioxide emission then will increase 60%, in accordance with the International Energy Agency's baseline scenario[*] pattern. Half of the increase will be caused by coal! It can be seen from this that coal is not the ideal method to resolve the energy problem; instead, it will bring a real nightmare to humankind.

---

[*] The International Energy Agency has formulated a baseline scenario plan for the future energy situation in accordance with the trend in 2006; the baseline scenario is not a prediction but an extrapolation of future energy trends based on present trends if no initiative measures are taken before 2030.

The present difficulty does not lie in how to eliminate carbon dioxide, which can be done using technology; rather it lies in how to reduce the cost to an acceptable extent and store carbon dioxide safely. The International Energy Agency deems that the reasonable cost of capturing carbon dioxide is 15–30 USD/ton; adding transportation (8 USD/ton) and storage (1–8 USD/ton), the final cost is not a small number.

But, even though a solution that is technologically and economically feasible can be found, we cannot be sure that climate disaster is averted. Clean coal power stations and carbon storage cannot be put into use until at least 2020–2025. It is necessary to digest and analyze experiment results conducted over the past 10 years. So, coal power stations will still emit carbon dioxide for now.

The Twin Brother of Petroleum—Coal

Coal is no doubt a commodity without charm. It is dirty, out of date, black, cheap, and even more inferior when compared with petroleum. Petroleum is a close relative of coal, but more dazzling and tactful. Petroleum takes adventurers, jet plane travelers, and international schemers to beautiful stages to fulfill their wonderful dreams. It has created a lot of extremely rich famous families and celebrities from Rockefeller to chieftains in the Middle East that people love and hate. Mining petroleum has led a few to get rich quick. The fortune tends to come not from hard work but from incredible good luck.

When we think of coal, we think of poverty not affluence. We immediately get a mental picture of miners covered with coal ash, dragging their feet in mines, eking a meager salary from oppressive companies, and supporting their poor and hopeless families with their hard dirty work. Coal is still regarded as valueless long after it has become part of our daily life. Petroleum is regarded as a symbol of luck and affluence, but coal is regarded as something disappointing. In reality, humble coal should be respected as a fossil; it is indeed a kind of fossil. Coal has existed before mammals appeared. At that time, marshes and forests could be seen everywhere and strange trees and huge ferns, which were "monsters in the plant kingdom," grew in them; coal was a molecule of the forests.

Most coal beds were already formed when the first batch of animals left the sea for land. Coal had witnessed the way for animals evolving from the sea to land and protected them to finish their evolution—those creatures that had dominated the earth before their advent were all extinct and coal is their highly concentrated remains. That we live comfortable lives is proof that the ecological environment gradually evolved toward the direction of being helpful to humankind. And those creatures all played important roles. If coal were not so energy rich, it is not difficult to view it in museums where it would be exhibited together with dinosaur bones.

In the past several centuries, coal accompanied us in our industrial revolution, laid the foundation for contemporary civilization, and finally gave us strength to build a world perhaps no longer needing coal. Hence, it deserves humankind's careful preservation in civilized museums; though it has a lot of defects, it has made great contributions to humankind.[*]

10.3 Renewable Energy Beset with Difficulties

The technological progress of developing and using renewable energy has been very rapid in recent years. Facing the worrying global climate warming, worse environment damage, exhausted petroleum resources, and limited coal reserves, we can only regard renewable energy as a life-saver that can revitalize humankind. Renewable energy neither releases carbon dioxide nor do its resources get exhausted. But, in fact, it is not as green as it looks and cannot meet the huge energy demand of the future world because its efficiency of use is still too low. Whether the prospect of depending only on renewable energy is bright or not is not clear.

The growing rate of the use of wind energy and solar energy—the two banners of renewable energy, respectively—reached 41% and 28% in the last 25 years (see Figure 10.2).[†] An incomparably bright future seems at hand. It seems that we can believe that renewable energy can meet most of the future energy demand in accordance with the current development trend; and its incomparably huge "reserves" further strengthen this confidence. The energy that the sun ceaselessly supplies to the earth, in addition to being used as solar energy, still can be converted to limitless wind energy, ocean currents, river currents, plant materials, etc. Some people believe that as long as energy saving measures are taken and economic support is provided to the renewable energy industry, the threat caused by the exhaustion of hydrocarbon fuels and global climate warming can be eliminated.

Unfortunately, the current situation is not so rosy. Although renewable energy has rich reserves, it also has the unignorable defect of too low energy density. Renewable energy of the same unit area can only provide meager energy, compared with nuclear energy and hydrocarbon fuel. If a nuclear power station occupying a

---

[*] See *Waste*, written by Barbara Fritz, translated by Shi Na, published by Zhongxin Press, 2005.
[†] See *BP Statistics of World Energy 2012.*

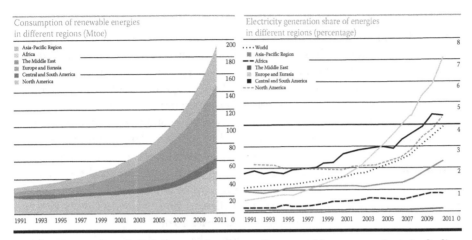

Figure 10.2 The electricity production of renewable energy (not including nuclear power and hydropower).

land of 10 ha can produce energy of 1.5 million W, a wind power station producing the same energy needs a land area of 18,700 ha.

Across the world, renewable energy has surprisingly great potential; but locally, it is relatively less popular and only generated by using solar panels, wind power generators, dams, and other relatively expensive energy-collecting equipment in an indirect manner. The already overcrowded earth can only provide limited space for producing an adequate supply of power, so the development of renewable energy is certain to conflict with the ecological system, arable land, and human living space. That the ecological system (such as forests) is very valuable is common knowledge; arable land is indispensable for human survival, and the reduction of human living space will bring more pressure to the development of renewable energy.

The International Energy Agency predicts that energy demand will still grow at a rate of 1.8%/year to satisfy the demand of global improvement of living standards, especially that of developing countries' residents. This means that the world's total energy production will go up from the current 11.4 billion tons of petroleum equivalent to 17.7 tons of petroleum equivalent by 2030, which requires that we increase the total energy production by 60% within the next generation while reducing the emission of carbon dioxide by 15%. Theoretically, renewable energy can finish the task. Energy production facilities of 0.5 billion tons of petroleum equivalent will be built and put into production each year from now on. As long as about 75% (about 0.4 billion tons of petroleum equivalent each year) of them adopt renewable energy, the need of controlling global warming will be met. In fact, however, achieving the goal is filled with great challenges.

Wind power and solar power generation have undergone rapid development lately. But among the types of renewable energy, hydropower occupies the most share, accounting for 16% of world electricity production; biomass power generation accounts for 1%; and geothermal power generation, solar power generation, and wind power generation account for only 0.7%. So, the starting point is too low. Production will be sure to slow down when a certain level is reached; this is in accordance with industrial law. Wind power generation already lags far behind in large-scale energy supply, and the supply shortage of silicon—the material of solar panels—will increase the cost of power generation significantly.

The greater problem wind and solar power generation face is that we cannot guarantee the constant supply of energy and cannot store large amount of electricity. To guarantee the safe operation of power networks, the energy of intermittent power generation can only account for about one-third of the total power, wind-driven generators or solar generators can only supply one-fifth of its installed power on average each year, and the electricity (kilowatt hour) provided by wind power generation network each year can only reach over 10%.

Solar power generation needs to overcome the great obstacle of integration into the power network, and its cost is 5–10 times higher than that of wind power generation. Although producers guarantee that the price of solar power will fall substantially, it is not going to come true in the near future. Besides, although the application of solar power generation technology in the nonnetwork arena is very promising, it can supply power to small power electric appliances and can let 1.6 billion people living in tropical regions to overcome the predicament of no electricity—it is like a hero with no place to display his prowess in developed countries where almost every family has access to electricity. These countries happen to be the main emitters of carbon dioxide. Hence, one can see that the future of solar power generation is similarly uncertain.

Maybe, these circumstances are deeply disappointing. In addition to the restriction by its own defects, the research for renewable energy has fallen behind the energy demand growth, it desperately needs a breakthrough on its technique innovation. Greater efforts still need to be made if it is to carry out the hard task of providing one-third of the total energy supply by 2030 and resolving the problem of global warming.

Wind Is Hard to Tame

On September 18, 2012, the China Report of Wind Power Development 2012 pointed out that in 2011, while new markets for wind power were developed, they faced problems of grid connection and periods of inactivity due to reduced wind. In 2017, the key to developing wind power generation in China was the achievement of rapid development in both old and new markets.

In 2011, the grid connection problem in China was eased to a certain extent, and the wind power industry showed rapid development. But now, the problems of too much and too little power are restricting power and curtailed wind power is very serious, and the development of wind power can be described as a "wind difficult to tame."

By the end of 2011, the accumulated installed capacity of wind power in China was 62,360 MW, grid-connected capacity 47,840 MW, and grid-connected rate 76.7%, a slight increase compared with 69.9% in 2010. In 2010, curtailed wind power was more than 10 billion kWh, and its rate was over 12%, equivalent to a loss of 3.3 million tons of standard coal, or equivalent to emitting 10 million tons of carbon dioxide to the atmosphere, leading to an obvious reduction of utilization hours of wind generators. In 2011, the number of grid-connected wind generators increased greatly, but their utilization hours on average were 1903 hours, a reduction of 144 hours compared with 2010. The regions where the phenomenon of reduced wind power was most serious were the "three norths" (northeast, north, and northwest China). Wind power enterprises' loss (excluding carbon trade income) caused by restricting power and curtailed wind power totaled over 5 billion Renminbi, accounting for about 50% of the profit of the wind power industry.

There is still a very broad scope for developing wind power given the present difficulty of harnessing the wind. The report predicts that the accumulated installed capacity of wind power in China will be 200–300 GW by 2020; more than 400 GW by 2030. At that time, wind power in China will account for about 8.4% of the national total electricity production and about 15% of the power source structure. The report also proposes that China should further make clear the local government's responsibilities and obligations regarding renewable energy, stick to the idea of giving the same importance to concentrated and distributed development, establish a quick-respond system to answer the power grid enterprise's needs and strengthen grid construction and control while carrying out the quota system of new energy.[*]

---

[*] See "China Report of Wind Power Development 2012," Li Junfeng, et al., China Environmental Science Press, 2012.

Chapter 11

Energy Disputes: Blasting Fuse of Future Warfare

11.1 Global Energy Problem

Mankind's massive and high-intensity development and use of natural resources have brought unprecedented economic prosperity since the first Industrial Revolution and created a brilliant industrial civilization. Yet, as the constriction of the global energy supply becomes increasingly acute, the energy crisis that sweeps the globe brings about a series of related contrasts: The contrast between population growth and resource supplies becomes sharp with each passing day; unreasonable resource development and use result in increasingly serious ecological environmental degradation; contending for energy resources leads to unstoppable warfare. If what energy caused at the beginning of the 20th century was only a few local problems—some industrial cities were shrouded in smoke all day; the British capital London became the foggy city that it is now—the crisis of energy at present has spread to every corner and nation on earth and has affected mankind's present and future.

Human consumption of natural resources has doubled since the Second World War. During 1901–1997, the prices of mined mineral resources in the world increased by nearly 10 times. Of them, in the last 20 years, it was 1.6 times what it was in the first 60 years. In 1950, the GDP per capita in each state was in direct proportion with the energy consumption per capita: The energy consumption per capita was below 1500 kg standard coal when the GDP per capita was 1000 USD; the energy consumption per capita was above 10,000 kg standard coal when the GDP per capita was 4000 USD.

The resources people live on have been continuously declining in the past several decades because of uncontrolled consumption (see Figure 11.1), especially the two energy crises that occurred at the end of the 1970s and the early 1980s, which swept the globe and shocked humanity. The price of crude oil at that time increased from less than 3 USD (in 1973) to 39 USD (in February 1981), leading to some countries enacting a policy of fuel rationing (see Figure 11.2).

At the same time, the trend of water and air pollution has intensified; local environmental deterioration has aggravated new global troubles. The population

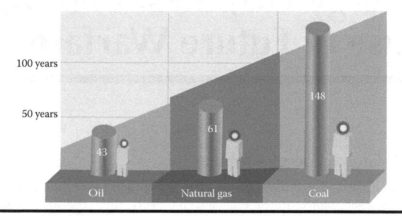

Figure 11.1 Predicted usable years of conventional fossil energy. (From BP, 2012.)

Figure 11.2 The motorcade outside a filling station in the state of Maryland, after the policy of fuel rationing was introduced. (Photo by Warren K. Leffler in 1979.)

growth rate is too high—world population has broken through the 7-billion mark, increasing by nearly two times from what it was in 1950. The pressure of rapid agricultural and industrial development has pushed aside other species and brought them to the brink of extinction. Excessive demand has led to the rapid erosion of the soil that mankind relies on for existence; this erosion has lowered the bearing ability of the earth itself and changed the atmospheric environment of the earth.

The energy problem has gone through a historical process of gradual development—a product of the excessive consumption of natural resources due to industrialization, it has become a global problem spreading to each corner and nation on the world. Mankind's recognition of the energy problem has also gone through a series of historical changes. So far, we have reached a consensus on the resources and environment crisis caused by the energy problem. The energy problem is always closely linked with population, environment, economy, society, and other issues. Since entering the 21st century, the pressure of a rapid population increase and economic development is exceeding the limit that resources and environment can bear. What mankind is facing is a devastated and overburdened planet where natural resources are reducing rapidly, biological species are on the brink of extinction, mineral energy resources are becoming increasingly exhausted, mineral resources are in extremely short supply, and ocean health is seriously damaged. Our resource treasure house faces a great calamity—fresh water resources are seriously in short supply, forest resources are continuously reducing, loss of water and soil intensifies day by day, greenhouse effect is worse and worse, climate change is continuously abnormal, natural disasters frequently occur.

Will an energy crisis similar to that of the 20th century happen in this century? It is not possible to make an optimistic prediction from the perspective of energy structure, regional distribution, political environment, and other aspects. At present, petroleum accounts for 39% of the world energy consumption structure, and two-thirds of the petroleum resources are concentrated in the Persian Gulf region. Besides, the environmental problem resulting from the massive use of coal, petroleum, natural gas, and other fossil energy is increasingly becoming serious; they are the root cause of local air pollution, acid rain, greenhouse gas emission, and other regional environmental problems. To guarantee the normal operation of the world economy that totals 20 trillion USD, Earth emits 6 billion tons of carbon dioxide to the atmosphere; this has already become difficult for the planet to bear.

Energy resources are important for human life, and the energy problem affects society again and again. Several big leaps in human modern history all benefited from the development and use of energy, and several major global crises were all due to energy. In the modern world of economic globalization and political multi-polarization, to guarantee energy supply and establish an energy safety system have become the starting point and core issues of the energy strategy of all nations in the world.

The Three Petroleum Crises in History

The first petroleum crisis. The fourth Middle East War broke out in October 1973. To fight against Israel and its supporters, the Arabian members of OPEC* (Organization of Petroleum Exporting Countries) announced a reassessing of crude oil pricing in December of the same year and increased their standard price of crude oil from 3.011 USD per barrel to 10.651 USD per barrel; the oil price in the international market went up from 3 USD to 12 USD per barrel, an increase of four times. So, this was the first serious economic crisis after the Second World War; it lasted for 3 years, striking a heavy blow to the economy of many developed countries. American industrial production went down by 14%, GDP by 4.7%; Japanese industrial production fell by 20%, GDP by 7%; European GDP dropped by 2.5%. After the crisis, some developed countries led by the United States formed the International Energy Agency to cope with any future potential petroleum crises. The organization requires member countries to maintain reserves of imported crude oil equivalent to 90 days of the previous year.

The second petroleum crisis. At the end of 1978, Iran launched a revolution and started a war with Iraq, resulting in a sharp reduction of daily petroleum production and the second petroleum crisis, in which global petroleum production declined from 5.8 million barrels/day to below 1 million barrels/day. Lacking 5.6 million barrels/day, the international petroleum price surged from 13 USD per barrel in 1979 to 35 USD per barrel in 1980. The situation lasted over half a year, becoming the main reason of the all-round decline of the western economy at the end of the 1970s; American GDP also fell by 3%.

The third petroleum crisis. The Gulf War of 1990 directly caused the third worldwide economic crisis. The international oil price soared from 14 USD per barrel to 42 USD within 3 months due to the supply interruption of crude oil from Iraq. The rate of progress of the U.S. economy declined quickly in the third quarter of 1990, resulting in the global GDP increase rate falling by 2% in 1991. The International Energy Agency started an emergency plan soon afterwards; 2.5 million barrels of reserved crude oil was put into the

---

* OPEC (Organization of Petroleum Exporting Countries) was established in 1960 for the purpose of coordinating and unifying the petroleum policies of the member countries to safeguard their respective and common interests.

market every day, the oil price slumped over 10 USD per barrel in a day, OPEC also rapidly increased production. Hence, the high oil price this time did not last long, having less of an effect on the world economy compared with the previous two crises.

The three petroleum crises had many common features: They were all because OPEC member countries' supply sharply dropped, making the market fall into the crisis of losing balance between supply and demand; all these crises landed a heavy blow to the world economy. Besides these crises, the international oil price also skyrocketed in 2000 when the global economic recovery began, after the 9/11 attacks in 2001, and in 2003 when the U.S. started a war against Iraq; but all these had little effect on the world economy at that time.[*]

11.2 The Political Attribute of Energy

Energy as an important strategic material needed for human survival and social development no doubt has a very clear political attribute. To take petroleum, the most important energy source, as an example, the competition surrounding it has never stopped since it was discovered. The reason lies not only in the material attribute of petroleum but also in its political attribute. Powerful nations all have had rich petroleum resources since modern times. In the world, the country that has more power and petroleum has the right of initiative in the strategy of international relationships.

Generally speaking, petroleum as a material resource by itself has no political attribute, but we cannot say that the development and use of petroleum resources and other problems are not related to politics. Its indispensability to modern society, plus regional partition of supply and consumption, deepen the geopolitical competition of petroleum-concentrated, countries and regions. So the prospecting for petroleum and its development, use, trade, transportation, possession, etc. lead to a lot of important political events.

The political role of petroleum is mainly that its consumption directly affects a country's economic development rate. Generally speaking, economic growth has a positive correlation with petroleum consumption, providing that the petroleum price is stable; on the contrary, economic development is affected when there is a short supply of petroleum. Petroleum as a kind of important strategic and civil material can make a contribution to the development of world economy. It can become the envoy to promote international peaceful communication, but it can also be the

---

[*] See *Contemporary World Economy and Politics*, Wei Ling, Sino-Culture Press, 2008.

trigger of wars and can even lead to a nation's rise or fall during war. Therefore, all countries try their best to control petroleum resources as much as they can. Whoever controls rights to petroleum holds' the advantage in politics, economy, and warfare.

Petroleum is closely linked with international politics and is often used as a means of achieving certain objectives in international political struggles. The Reagan administration sent people to major oil producers in the Middle East and Europe to lobby for increased crude oil production in large quantities at the beginning of the 1980s when it learned that the Soviet Union had gained huge profits through selling great amounts of oil; the petroleum price in the world went down sharply at once. Hence, the Soviet Union's foreign exchange revenue dropped by a huge margin, its scores of large industrial projects were forced to be scrapped, its internal structure became worse with each passing day; therefore, one of the important reasons for the Soviet Union's collapse was the American "petroleum weapon." As the new century began, the United States spared no risk of following in Britain and the Soviet Union's steps and falling into the "Afghanistan trap" by starting "the century's first war." Ostensibly, the war has ended with the victory of the western countries led by the United States, but the overt or covert war for petroleum resources in the region is far from over; it has just begun. America sent troops round by round in the Middle East and Middle Asia and even ignited the powder kegs of Iraq. Emerging Russia is fiercely contending with the U.S. for "oil pipelines" and "oil bowls" in the Middle East–Caspian Sea regions to achieve its own dream of becoming a powerful nation. On October 15, 2007, Russian President Putin, despite the rumor of an assassination plot on him, flew directly to the Iranian capital Tehran to hold talks with the Iranian president Mahmoud Ahmadinejad after visiting Germany. He took part in the summit of the countries along the Caspian Sea coast held the next day, the first visit by a Russian leader to Iran 60 years after the Second World War. Putin's desire to have closer ties to the Middle East–Caspian Sea regions, where there is abundant petroleum and natural gas resources, was no doubt the intent of the whole exercise. At present, Russia has risen again due to petroleum and has regained the status of a powerful nation—it is by no means a nobody. As global petroleum shortages become apparent, the scramble for petroleum is predicted to become more intense.

So far, a nation's policy is still greatly affected by energy, especially petroleum, no matter whether it is a world power or a developing country. Petroleum itself is just a material resource without strategic meaning, but for a nation, how much it has or how much ability it has to gain it is a major issue with strategic meaning. It is almost impossible for a nation to achieve economic development, political stability, and military safety without a sufficient, stable petroleum supply at reasonable prices. For nations with a high dependency on foreign petroleum, it is inevitable that they be involved in international affairs related to petroleum. So, we have to admit that petroleum is the greatest political weapon. Petroleum, as the major factor in international politics, affects the foreign policies and strategies of all countries of the world; it changes world political and international relationship patterns.

The petroleum shortage has not been eased; it is increasingly becoming more intense since the 21st century. Petroleum has become so important that people will attain it by hook or crook, at all costs. Today's many disputes are all related to petroleum; Africa, South America and the Middle East are all in turbulent situations because of petroleum.

The political attribute of petroleum as the most representative energy is evident. Other sources of energy, such as hydroelectricity from rivers flowing through several countries, the still-growing ocean energy, and nuclear energy, also have an important effect—directly or indirectly, they affect the world political situation and are branded with an indelible political mark. We can infer that those who possess energy influence the world.

Petroleum Influences Politics

It is generally assumed that the search for energy is Britain and America's fundamental principle of foreign policy. Some people point out that the British release of the Libyan Abdelbaset Ali al-Megrahi, who had been sentenced for the bombing of an American jetliner over Lockerbie, was an attempt to obtain Libyan petroleum and natural gas resources. The British government of course did its best to eliminate this view, asserting that the prisoner had been handed over to Libya through "extensive negotiation" and in the light of Britain's "huge interests." Although the word "petroleum" had no place in the plea, later, the country had to admit that trade and petroleum interests had played a "very important role" in Britain's desire for "reaccepting" Libya.

To be sure, petroleum is not the only interest that Britain has in Libya, but looking for safer and more multivariate energy sources is no doubt more and more important for Britain's foreign policies. British energy reserves in the North Sea region are gradually being depleted, so the nation is worried about the fast approaching energy crisis. Libya seems to be a promising and potential petroleum and natural gas supplier and is especially open to foreign petroleum companies. Both British Petroleum Co. Ltd. and Royal Dutch Shell that are respectively listed as the second and the third in market value in the London Stock Exchange have signed prospecting contracts with Libya.

The relationship between Britain and Libya is just one example of a general phenomenon; energy is at the core of many important international political problems. This is because there is not an economy in the world—U.S., Japan, or Europe—whose petroleum and natural gas resources are self-sufficient. The global demand for petroleum is steadily rising, and all major economies are competing for energy resources.

The American scholar Michael T. Klare* points out in his book *Rising Powers, Shrinking Planet*, "a fierce competition will rise in the expanding group of energy consumers in the world of rising powers and exhausting resources." The U.S. national intelligence director, Dennis Blair, also recommends this book.

There are many incidents that showcase the fierce competition for energy that has begun to affect the foreign policies of major world powers. The intense situation between Russia and the European Union is mainly because the latter increasingly relies on the former's energy supply. Although western powers are likely to try to intensify sanction on Iran because of its nuclear program, India is prudent because Iran is its major energy supplier, and Indians still hope to build a natural gas pipeline to carry Iranian natural gas to their domestic market.

Major energy-demanding nations hope to sign energy agreements if possible. At present, the international society is also interested in Africa because its petroleum resources increase with each passing day; Angola, Nigeria, Congo, and other countries all have become hotspots of international concern. For example, leaders of every country are visiting Angola. The then U.S. Secretary of State Hillary Clinton visited Angola in August 2009. Russian President Dmitry Medvedev also visited it in June 2009. Besides Angola, Brazil has become more and more popular since a new big offshore petroleum field was discovered there.

There were no doubt many reasons for America to invade Iraq, but the former chairman of the Federal Reserve of the United States, Alan Greenspan, said in his memoir, it is "politically inconvenient to admit the fact everyone knows: the Iraqi war is mainly related to petroleum."

No matter whether Greenspan's conclusion is right or not, it is certain that American leaders are highly worried about the obtainability and price of petroleum. The stagflation caused by the two petroleum crises in the 1970s had perplexed Europe and America in those 10 years. Compared to those dismal

---

* Michael T. Klare, an American expert in world peace and safety, wrote *Resource Wars* and *Blood and Oil*, etc.

years, the long-term prosperity during the Reagan and Clinton administrations was all supported by low oil prices. The Soviet Union's collapse was closely related to the fall of oil price in the 1980s; Russia became richer and more confident because of the rise of oil price in the past 10 years.

The famous historian of the petroleum industry, Daniel Yergin,* say, "it is petroleum that makes our living places and life style become possible... Petroleum and natural gas are also the key materials of chemical fertilizer, on which the whole world's agriculture relies; petroleum lets us transport our food to the big cities in the world that can not be self-sufficient."

Politicians know that if fuel prices soars or energy is scarce, they will be punished by their voters. But they know, at the same time, that if looking for petroleum is pursued as a matter of foreign policy, they will likely be condemned for greed and immorality. Western politicians behave like innocent children with their voters when the energy safety problem is involved, but act truly like adults in private.

11.3 Energy and War

If one say that wars are mostly due to energy and that we cannot do without it, nobody will oppose this statement. Ancient wars were usually closely related to biomass energy and animal power energy. Ancient people said, "Grain and fodder should go before troops and horses," which means that animal power of horses and other domestic animals was used for transporting grain and fodder for troops. The horse-driven chariots were important for wars that occurred in the spring and autumn period during the Warring States period in China; chariots were also important in Greece.† A "state with 1000 chariots" and a "monarch having 10,000 chariots" indicates the importance of chariots to nations. That Genghis Khan could lead Mongol cavalry to sweep Eurasia showed the power of horses; the purpose of war those days was to obtain grain, fodder, livestock, water resources, and other energy resources.

Today, the fighting capacity of troops has obviously improved after cars, tanks, armored vehicles, airplanes, and warships replaced domestic animals. The key to victory in a war lies with the army that has the stronger fighting capacity;

---

* Daniel Yergin, an American energy expert, chairman of Cambridge Energy Research Associates, wrote *The Prize: The Epic Quest for Oil, Money, and Power* and other books.
† In ancient China, the chariot was driven by four horses, the power of vassal states was measured by the number of the chariots. A state with 1000 chariots was a medium state; a monarch having 10,000 chariots was the monarch of a powerful state.

the fighting capacity lies in firepower and maneuverability, and maneuverability means the ability to put troops quickly to battle. At the same time, mutual coordination between troops and weapons is very important. Those who more quickly supplement ammunition and give the enemy a fatal blow can better preserve themselves and realize their strategic goal. The spatial span of modern battle is no longer what it was; an enemy's military targets can be attacked from a range of 10,000 km. For example, during the Afghanistan and Iraqi wars, American B-2 stealth bombers could fly to the Gulf region and to Afghanistan to attack there after starting from America and after air refueling. Modern wars have taken on an entirely new look. The resources that they rely on and contend for have also been mainly petroleum.

The history of the past 100 years is the history of plundering and controlling the petroleum reserves of the world; almost each war, conflict and turbulence was directly or indirectly related to petroleum. Especially after the 1950s when the world entered the "petroleum times," there was an endless stream of conflicts and wars in the world caused by contending for petroleum resources. Petroleum is like an opened Pandora's box, the contention it caused was never as serious as it is today.

It was at the beginning of the 20th century that petroleum became the major strategic resource for which western industrial countries contended. The Allied Nations led by Germany and the Entente Countries with Britain, France, and Russia as their core fought for strategically important areas and petroleum resources in the First World War (1914–1918). Both parties at war consumed more than 13 million tons of fuel altogether, and the Entente Countries defeated the Allied Nations relying on American and British petroleum. The British statesman George Nathaniel Curzon's* utterance at a petroleum conference after the war became famous. He said, "the Entente Countries drifted toward victory on the waves of petroleum." Petroleum had become the supportive strength of modern wars.

In addition to politics, economy, and military affairs, petroleum has also contributed to terrorism; terrorists took advantage of the fact that transportation was driven by petroleum; for example, 19 terrorists in the 9/11 event hijacked an airplane to destroy New York's World Trade Center. The event shocked the world, and the United States started 21st century's first antiterrorism war afterwards, whose reasons, purpose, and course were all closely linked to petroleum.

In the present world, the regions that are always turbulent and where there are frequent wars and conflicts are the regions with rich petroleum reserves. The number of armed conflicts and wars that have taken place in the Middle East since the Second World War ended are more than 30. The massive wars that America has started since entering the 21st century, such as the Afghanistan War and the Iraqi War, are all directly and closely linked with the rich petroleum resources in

---

* George Nathaniel Curzon was British Secretary of State for Foreign Affairs after the First World War.

the warring places, such as the Middle East and Middle Asia. Whether the petroleum resources were obtained by starting the war or controlled after it, reflected the interests of petroleum groups, which sought military protection during wars and in emergencies. Petroleum is also an important means by which the U.S. competes with its rivals; it played an important role in the Cold War when the U.S. secretly and indirectly battled with the Soviet Union. Therefore, petroleum has an important effect on world politics and economy; it not only affects victory or defeat in war, but it also affects the rise and fall of nations.

Iraq, the place where the Babylonians created the Hanging Gardens, a wonder of the ancient world, is abundant in petroleum resources. It is listed as having the third largest petroleum reserve in the world, and it is blessed with the most favorable natural conditions among all countries in the Middle East. But because of this, Saddam Hussein overestimated his nation's strength and made a mistake in national strategic decisions when he started a war against Iran. The war lasted for 8 years and caused destruction on both sides, killed and injured millions of people, resulted in the loss of a large petroleum fortune, and greatly affected the economy of the two nations. Iraq is still in turbulence and conflicts.

In addition, because world powers and internal influences contend for petroleum, wars and conflicts frequently occur in Africa, South America, and Middle Asia, especially the countries in Africa that produce petroleum most, such as Nigeria, Angola, and Libya. Angola is one of the main countries on the west coast of Africa; its civil war has lasted more than 20 years, with the contending powers fighting for state power and petroleum and diamonds. In Sudan, the people from the northern part mainly believe in Islam, and people from the southern part mainly believe in Christianity. Conflicts between the two parts frequently occur because of religious opposition and petroleum resource distribution. South America, Columbia, Venezuela, Mexico, and other major petroleum producers are similarly facing turbulent times.

Energy not only creates a brilliant civilization but also causes brutal wars. Upon final analysis, civilization or war lies in the hands of those who control energy. How to reasonably distribute and use energy, achieve common prosperity and lasting progress of civilization, and avoid the pain brought by wars and turbulence is an important issue. We who have entered the new century and hope to start a new civilization must face this issue squarely, rethink profoundly, and settle on a solution.

Global Four "Powder Kegs" of Energy

In 2009, Russia clearly pointed out in its Strategy of Energy that contending for energy will be the trigger of future wars. In fact, the competition among all countries for the control of the Middle East, the continental shelf of the Barents Sea, the Arctic, the Caspian Sea, Middle Asia, and other important places producing energy never stops and has created four powder kegs of energy.

China and Japan have always disputed over the East China Sea. In 1968, a report by the UN Economic Commission for Asia and the Far East (ECAFE) pointed out that there were 20 billion m^3 of natural gas and a large amount of petroleum in the East China Sea between China and Japan. After that, the dispute between China and Japan about the demarcation problem of the continental shelf of the East China Sea became more serious. Its root cause is the division of the boundary of the exclusive economic zone, and its focus is concentrated on the sovereignty of the Diaoyu Island and its affiliated islands, which are situated at the edge of the continental shelf of the East China Sea of China; Diaoyu Island and its affiliated islands were first discovered and named by China. Japan unilaterally began to prospect them and attempted to occupy them at the end of the 1960s after it was thought that there were probably large amounts of petroleum and natural gas in the vicinity of the islands. The Japanese government unilaterally announced that it would "purchase" the Diaoyu Island and its affiliated islands to nationalize them on September 10, 2012, which seriously violated China's territorial sovereignty. So far, the war cloud has not dispersed.

The sovereignty dispute of the Arctic. There are extremely rich natural resources in the Arctic. In addition to the renewable natural resources, such as fishery, water power, and wind power, there are still nonrenewable mineral resources, such as petroleum, natural gas, copper, cobalt, nickel, lead, zinc, gold, silver, diamond, asbestos, and rare elements. So the Arctic is called "the second Middle East." Besides, with global climate warming and acceleration of ice and snow melting, the northwest channel of the Arctic connecting the Atlantic and the Pacific will be open and the route of travel from Asia to Europe will be shorter by over 10,000 miles, which can not only save freight costs but is also of significant military meaning. So far, Denmark, Iceland, Norway, Sweden, and Finland in Northern Europe; the United States of America; Canada; and Russia have all claimed their sovereignty over the Arctic because there is internationally no law that clearly stipulates its ownership. The sovereignty dispute of the region is likely to cause a war.

America and Iran pinch each other's "throat of petroleum." Iran has abundant reserves of petroleum and natural gas. By the end of 2006, its proven petroleum reserves were 138.4 billion barrels, and it had 27.51 trillion m^3 of natural gas, second only to Russia. Petroleum is Iran's economic lifeline, and its receipts account for more than 85% of all Iranian foreign exchange receipts. Hence, Iran has become the second largest petroleum-exporting country in OPEC. It sent troops to occupy Abu Moussa Island and Greater and Lesser Tunbs near the Strait of Hormuz, which originally belonged to British protected territory. The three islands are situated at the Arabian Gulf, near several main petroleum fields of strategic importance. Iran basically can blockade the Strait of Hormuz and control the shipping channel of the region in reality. America deploys a great number of troops in the Arabian Gulf and surrounding regions. The headquarters of its fifth fleet is in Bahrain, and there are about 30 warships of American troops and allied forces in the region. That American forces eagerly entered the Middle East is because of its important military location and energy resources. The Strait of Hormuz between Iran and Oman has become the land that America urgently needs to protect. A blockade by Iran would be a heavy blow to America.

Small countries in Africa fall into turmoil due to petroleum. Equatorial Guinea and other countries have discovered oil fields one after another since the 1990s, and the Gulf of Guinea has become an important energy base with total oil reserves likely to be more than 24 billion barrels; it is promising to be the new Middle East. But the coastline of Angola, Cameroon, Gabon, Equatorial Guinea, Nigeria, and the Democratic Republic of the Congo around the Gulf of Guinea is still divided. These countries can still more or less deal peacefully with their disputes at present; although Equatorial Guinea and Gabon have been entangled in a land dispute over a handful of small islands occupied in 1972, and there are frequently violent conflicts in Nigeria and coups in several island states. Besides, the West African government is seriously corrupted, and the fortune that some countries seize from petroleum is not used for improving their people's living standard. All these are likely to cause dissatisfaction among the people. The situation in these regions is extremely unstable.

Infinite Power: The Dream We Chase

At each turning point of civilization, human beings seem to fall into dilemmas but they always get out of it. Since we entered modern society, facing many difficulties and challenges, we keep seeking the dream of infinite power. Although the road ahead is still filled with thistles and thorns, satisfying progress has been made in technical innovation. Let us look for the sunshine through the clouds. Systematic reform has been fruitful and has paved the way for our march forward. Many difficulties have been eliminated through international cooperation; let us feel the strength of unity.

Chapter 12

Technical Innovation: The Sunshine Passing through the Clouds

12.1 The Improvement of Traditional Energy Techniques

Civilization cannot progress without energy and power; energy and power in the meantime cannot do without energy technology. The history of human civilization is, to some extent, the history of energy technology's progress. We have been innovating and making progress and making energy cleaner and more efficient over a long period of exploration and accumulation, whether it is drilling wood to make fire or using wind power, hydropower, thermal power, petroleum, nuclear power, etc. We can see how traditional energy technology has improved from the technological change of wind power utilization and thermal power generation.

Technology of wind power utilization. Wind power is a traditional source of power that people have used for ages. It has a history of thousands of years, beginning from when human beings used domestic animal power and hydropower in agriculture. Before the steam engine was invented, wind power was an important driving force. Sails enabled people to explore and travel across rivers, lakes, and seas, while windmills helped people draw water, irrigate fields, grind flour, saw wood, extract oil, and drain away excess water.

Wind-driven machinery was gradually eliminated because it could not compete with the steam engine, internal combustion engine, and electric motors of modern times. It was only in the middle of the 1970s, because of the scarcity of fossil fuel, that attention was paid to wind-power-utilizing technology again. It was improved and developed, especially wind power generation technology, which has become the most important representative of renewable energy.

Denmark developed a wind power generator and built the first wind power station at the end of the 19th century. In 1931, the Soviet Union constructed the biggest wind power generator set (30 kW) in the world at that time using propeller type blades. Developed countries made significant progress in developing wind power generator sets in the 1980s; horizontal shaft wind power generator sets with a unit capacity of 3.2 MW and vertical shaft wind power generator sets of 4.0 MW were developed. Wind generator sets with a unit capacity of 100–200 kW dominated the middle and large sized wind power stations in the 1990s. The installed capacity of wind power generation in the world has rapidly increased in recent years, and America, Germany, France, Denmark, and China (see Figure 12.1) have paid great attention to wind power generation. The total installed capacity of wind power generation in the world had reached 282 million kW by the end of 2012, with an annual installed capacity more than 40 million kW. Wind power generation had become the fourth biggest electric source after hydroelectricity, coal power generation, and nuclear power generation.

The technology of wind power generation was further improved in the 21st century; new designs of wind wheels, adjustable rotors, direct driving, variable speed conversion system, power electronics, and high quality materials, etc., were used, and so the wind power generator unit capacity continues increasing—the format has diversified. Continuous improvement of manufacturing technology, utilization rate, and mature design technology can make a unit run 20 years or longer, even under unsteady and uneven load and poor climate. In the future, in the technology

Figure 12.1   Wind power station, Daban Town, Xinjiang, China.

of wind power generation, operating reliability and stability will be continuously enhanced and equipment investment and generation cost reduced, so the unit capacity of the wind power generator sets, which is currently in the range of hundreds of kilowatts, will go up to the megawatt range. High pylons made of lightweight materials and advanced wings will be used, and freestanding and multipart floating wind wheels will be developed. Wind power generation with clean and safe characteristics could meet the demand of globally sustainable development; its competitiveness will continue to improve, and its share in global electricity generation will increase as well. It has become one of the renewable energy resources that has developed the fastest and whose cost is the lowest; it will effectively reduce the world's dependency on fossil fuel and reduce the emission of carbon dioxide.*

The technology of thermal power generation. Thermal power generation is the process of using thermal energy produced by coal, petroleum, and natural gas to heat water to produce high-pressure steam that can drive steam turbine generators to produce electricity. It is the oldest and most important among all power generation methods. The earliest thermal power generator was in the thermal power plant of the Paris North Railway Station in 1875. Thermal power entered a phase of rapid development after social electrification period during the 1930s owing to the improvement of generator- and turbine-making technologies, innovation of transmission and distribution technology (especially the appearance of power system).

The large amount of exhaust gases, ashes, and noise produced in the course of thermal power generation led to pollution of the environment. In recent years, technologies have been improved through many approaches, such as adopting desulfurization and denitrification devices, carbon dioxide capture and storage, cofiring of biomass and coal, addition of man-made methane to natural gas, conversion of coal into natural gas, etc., which have resulted in many sophisticated systems like the natural gas combined cycle system, coal-fired power plant gasification cycle system, fluidized bed combustion, integrated gasification combined cycle system (see Figure 12.2),† fuel cells, clean coal, and other advanced technologies.

Clean coal technology is the generic term given to a series of new technologies of processing, combustion, conversion, and pollutant control for improving coal use efficiency in the whole process from developing coal to using it, making it a kind of energy that can be used to the hilt; it includes clean production technology, clean processing technology, high-efficiency clean conversion technology, high-efficiency clean combustion technology and coal pollution and emission control. Clean coal power generation includes the following key technologies: first is washing of coal

---

* See *The Future of Energy Use*, Phil O'Keefe, Petroleum Industry Press, 2011; *Modern Energy and Power Generation Technology*, Xing Yunmin and Tao Yonghong, Xidian University Publishing House, 2007; *Energy Technology Perspectives*, International Energy Agency, translated by Zhang Eling, et al., Tsinghua University Press, 2009.

† IGCC (Integrated gasification combined cycle) refers to an advanced power system that combines coal gasification technology and the high-efficiency combined cycle. It consists of coal gasification and gas–steam combined cycle power generation.

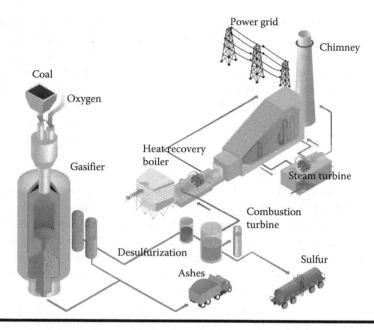

Figure 12.2   High-efficiency environmental IGCC power station.

to reduce impurities. Up to 50–70% of ash content and even 30%–40% of sulfur content can be generally eliminated. Second is cleaning in the process of coal combustion, which is the core of clean coal power generation. It uses low nitrogen oxide combustion technology to improve combustion and lower pollutant emission also it is a technical measure that is both economical and easy to set up. The third key technology is gas cleaning, namely, gas desulfurization to reduce the emission of the sulfur dioxide of the boilers in power stations, which is the technology of controlling the emission of the sulfur oxide—it is most widely used in the world. The fourth key technology is conversion, mainly coal gasification and liquefying, which can effectively enhance the utilization rate of coal heat energy and reduce the pollution caused by directly burning coal. Clean coal technology greatly reduces the harm coal has on the environment and humans while developing and using coal and is an important improvement and development of thermal power generation. The gas and liquid fuel obtained from the traditional nonclean energy source—coal—greatly raises the conversion rate in thermal power generation.

There is another gas turbine power generation technology that has made power generation very efficient. Fuel and compressed air are put into the combustion room together; the heat of this mixed combustion is not used for heating steam but for making gas with a certain speed and direction to enter the gas turbine and drive its rotors to rotate at high speed to drive the generator and produce electricity. Compared with the steam turbine, the gas turbine that directly uses the pressure and temperature of gas simplifies the conversion process of energy. It was Germany that invented the method of "gas turbine–steam turbine combined cycle power generation," which greatly enhanced efficiency by using the energy of one combustion to generate electricity twice. At present, its heat utilization rate has reached 47% and is likely to be enhanced further.

The Wisdom of Constructing an Energy Efficiency Building

Energy saving does not mean we need to decrease our quality of life or use less energy; it means we need to enhance end-use energy efficiency, make better use of energy by using technology more reasonably and more efficiently. We need to improve and develop traditional energy. The technology of building energy efficiency is closely related to human life. It refers to setting up energy efficiency standards; adopting energy efficiency technology, process, equipment, materials, and products; intensifying the operation management of the energy system for buildings in the course of building plan, design, construction, transformation, and use; and reducing energy supply, air conditioner refrigeration and heating, and energy consumption of lighting and hot water supply without compromising indoor environment quality. The Shanghai World Expo Center 2010 (see Figure 12.3) can be used as a model of a public building that uses energy-efficiency technology.

From the beginning, the Center set up many domestic and international standards related to energy efficiency and

Figure 12.3   An energy-efficient building: The Shanghai World Expo Center 2010.

the environment and strictly implemented them. They made overall arrangements for resources and energy efficiency, recycled and reused in accordance with the 3Rs (reduce, reuse and recycle) design principle, and started with saving energy, materials, and land so as to reduce resources and energy consumption, pollutant emission, the effect of the building on the environment, thus really enabling embodiment of the "Better City, Better Life" theme of the world expo.

The total energy consumption of the Center was lower than 80% of the value required by national energy efficiency standards; the building energy saving rate was 62.8%, nontraditional water resources use rate was 61.3%, and renewable building material use rate was 28.9%. Each year, the Center saves an amount of energy that is equivalent to 2160 tons of coal equivalent (equivalent to the annual electricity consumption of over 10,000 residents in Shanghai), reduces carbon dioxide emission by 5600 tons, and saves running water of 160,000 tons (equivalent to the annual water consumption of over 10,000 residents in Shanghai). To realize these energy saving targets, the Center relies on reducing building energy consumption and reducing emission to its maximum.

The north and south exterior walls have been designed differently. Because glass walls are often used in urban public buildings, which will not allow the building to be naturally ventilated, the electricity consumption to run fans and pumps inside the buildings is huge; therefore, the energy consumption problem is becoming increasingly serious—it has become a worldwide challenge. To reduce the energy consumption of buildings, the Center used transparent glass curtain walls in the north exterior wall that is not exposed to the scorching sun in summer and there is just enough sunshine for use of daylight. With the world EXPO garden and Huangpu River nearby, visitors can fully enjoy the beautiful scene outside. The south exterior wall, which is exposed to plenty of sunshine, has a self-sunshade, vertical design, double-layer glass curtain wall in which there are wire meshes and inert gases that help in sun-shading and heat insulation; they can limit direct sunlight in the hot summer and reduce excessive heat from entering the rooms, thus reducing energy consumption and creating a comfortable indoor environment.

Solar energy is used to its maximum. To maximize saving energy and reduce emission, the Center adopts the technology of using solar energy. Five thousand three hundred and sixty conventional solar modules are installed on the top of the building, and 1064 photovoltaic sun-shading modules are installed on the south wall of the equipment room on the top of the building, with a total installed capacity of up to 1 MW, providing electricity to the grid by connecting to the city's power supply system, saving 357 tons of standard coal each year and reducing carbon dioxide emission by 950 tons. The photovoltaic sun-shading system is a new type of photovoltaic unit combining solar photovoltaic technology and traditional sun-shading devices. It can provide shade from the sun; avoid temperature rise of indoor air, wall, and floor; and effectively improve indoor comfort.

Use of river water and recycled rainwater. Taking advantage of being near the Huangpu River, the Center adopts new energy conversion technologies, using river water, and heat pump, ice storage, water storage, and rainwater collection systems. The river water source and heat pump system use the water in Huangpu River for refrigeration in summer and for heating in winter. Compared with gas heating, it can reduce energy consumption by 40%–60%, operational cost by 50%–70%, and carbon dioxide emission by 2660 tons, thereby saving 1000 tons of standard coal as it operates once each year.

Energy saving and emission reduction play significant roles; it can beautify buildings, help ease urban hot island effect, and save a great amount of resources.

Besides, the Center has also established a perfect and comprehensive rainwater control and use system, which can collect and use rain water of up to 30,000 m³ each year, accounting for over 14% of annual water consumption. Another nondrinking water collection and use system collects and uses nondrinking water of about 123,000 tons, accounting for about 58% of the annual water consumption. The Center uses water-saving sanitary hardware and fittings to save water and uses program-controlled grass land microirrigation systems, which save 50%–70% more water than flood irrigation and 15%–20% more water than spray irrigation.

12.2 The Exploration of New Energy Technologies

In addition to the improvement of traditional energy technologies, we have always proactively explored different kinds of new energy technologies. We have made significant progress in photovoltaic power generation, biomass power generation, hydrogen power generation, shale gas development, fuel cells, and other energy development and utilization technologies and have made important breakthroughs in compressed air, flywheel, and other energy-storing technologies; superconductors, ultrahigh-voltage, and other transmission technologies; intelligent grid; and other comprehensive use technologies.

Photovoltaic power generation. It uses the photoelectric effect of semiconductor materials to convert the light energy of the sun into electric energy. The solar cell made of semiconductors is the key element needed for photovoltaic power generation. It can produce voltage and current from the energy of the sun and convert solar energy into the electric energy needed in daily life (see Figure 12.4). Silicon, an abundant natural resource, is the main raw material of solar cells. It can be refined to a certain purity and a little amount of additive (such as boron and phosphor) is added to make silicon atoms produce charge and possess electric conductivity. Solar cells produce direct current like common alkaline batteries when they absorb sunlight. The current changes with changes in sunlight. As the photovoltaic material technology develops, different types of solar cells are possible: the ability of film photovoltaic modules to absorb sunlight is much stronger than that of crystalline materials, but their output power gradually declines; and extended-type integrated photocells can be used in space, and their efficiency is much higher than other cells. As organic polymeric material is being developed, more stable and lower cost solar cells can be obtained.

Biomass power generation. This is also a new trend of the future. It can be done in two ways. One way is for biomass to be directly burned to generate electricity, that is, it is directly burned to produce steam to drive generating equipment to produce

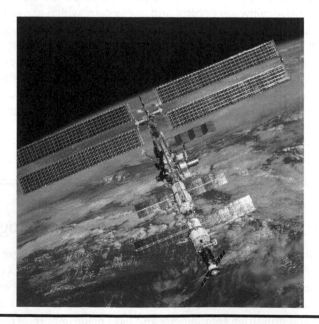

Figure 12.4 Photovoltaic cells on the International Space Station.

electricity, similar to the working principle of thermal power stations. This technology has been in use for a long time and has gradually spread; it is used quite often. The other technique is biomass gasification power generation; its basic working principle is that biomass is converted into flammable gas to drive gas-power-generating equipment to produce electricity. It can fully use biomass energy and leads to higher efficiency power generation. Therefore, it is the basic technology that is extensively used in the biomass industry. There are innumerable ideas in the development of biomass power generation technology, such as artificially propagated plants being used to produce biological diesel and even petroleum. The development of the field of "energy agriculture" perhaps will bring us an unexpected bonanza.

Hydrogen energy utilization. Hydrogen power generation is the technology that uses the chemical reaction of hydrogen and oxygen to generate power. It mainly has three ways: combustion (Simply burn the Hydrogen), chemical reaction to use Hydrogen to generate electricity and the nuclear energy released by hydrogen. Hydrogen fuel cells (see Figure 12.5) have been rapidly developed through the concerted efforts made by America, Japan, Germany, Britain, France, and other countries. It can make energy conversion efficiency reach 60%–80% without combustion and with less pollution and low noise, and its size can be flexibly set. Hydrogen fuel cells can be used not only for the power generation of massive fixed power stations but also miniaturized to become the mobile electric source providing strong power for cars and ships. Hydrogen cars are technologically feasible, its fuel use rate is higher than gasoline cars, and its combustion product is water, which does not harm the environment.

Shale gas development. Shale gas is the natural gas mined from shale beds. It is an important nonconventional natural gas. Its formation and concentration have their own unique features. It is present in the shale rock strata in basins that are relatively thick and widely distributed (see Figure 12.6). Compared with conventional natural gas, it is characterized by a long mining life and production cycle. Most of it is widely distributed, relatively thick, and generally contains gas, which makes its wells dependably produce gas for a long time.

With the use of new drilling technology and breakthroughs in fracturing technology, the gas in the underground shale strata has been massively developed since the 21st century, which has caused an energy industry revolution called the "shale gas revolution." The U.S. with advanced shale gas mining technology, has included it in their national strategic energy list and is massively and proactively developing it. Shale gas production U.S. has accounted for over 30% of its total natural gas production. The U.S. has become the biggest natural gas owner and producer. Its natural gas price has reduced by more than 80% since 2008, so it has thoroughly turned around the trend of energy self-sufficiency fall; its energy self-sufficiency rate has been gradually raised, reaching 81% in the first 10 months in 2011, which was the highest level since 1992.

Fuel cells. It is a kind of power-generation device that directly and effectively converts the chemical energy stored in fuel and oxidants into electricity. With the

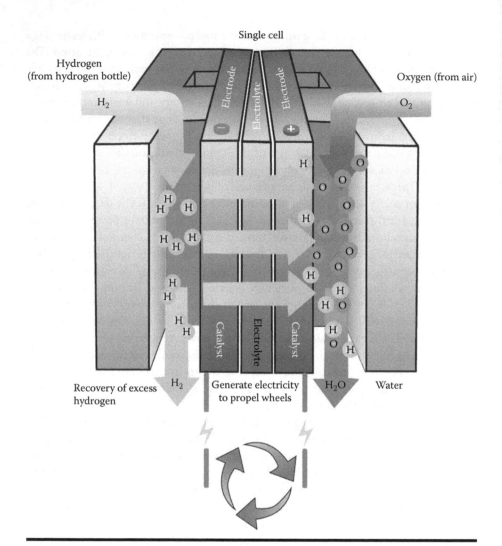

Figure 12.5 Principle of hydrogen fuel cell.

advantages of fuel diversification, clean exhaust, low noise, low pollution, high reliability, and ease of repair, it is regarded as one of the completely new, high-efficiency, energy-saving, and environmentally friendly power-generation techniques. Its power supply system mainly includes a fuel cell, a DC converter, an inverter, and an energy buffer.

Compressed air technology. The technology uses electric energy to compress air during low grid load to seal high pressure air in scraped mines, subsided sea bed air tanks, caves, and old or new oil–gas wells in order to release it during grid load peak to drive turbines for power generation. The compressed air is mainly used for power peak adjustment and system backup.

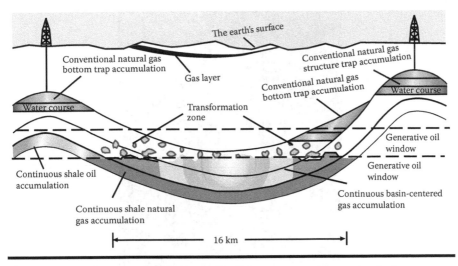

Figure 12.6   Diagram of shale gas and other energy resources in strata.

Flywheel technology. It is an important new mechanical energy-storing technology. It stores energy or momentum on high-speed rotating flywheel rotors, converting the stored energy from electric energy to mechanical energy and to electric energy again (see Figure 12.7). The flywheel consists of a high speed rotor, a bearing supporting the rotor, a high-speed reciprocal electric machine, and a control system. Its energy storage has many advantages, such as high density, big peak power, high conversion efficiency, and no pollution and charging and discharging time limit. Density is an important index of weighing energy storing flywheel. To raise rotating speed is the most effective means of achieving high density; to achieve high rotating speed, we should first resolve the problems of rotor support, high-speed driving, and materials and adopt magnetic suspension bearing and composite materials with high specific strength. Its disadvantage is that it has relatively low energy density and high cost in guaranteeing system safety. Its advantages are numerous. It is mainly used as a supplement of a battery system.*

Superconduction. It is the conducting phenomenon in which there is no resistance (zero resistance) at a certain temperature. An object that has this phenomenon is called a superconductor (see Figure 12.8).[†] In the 21st century, superconductor technology has shown the most potential of development in the power science field. Much progress has been made in its application in power

---

* See *Energy of Wisdom*, Wang Yi, et al., Tsinghua University Press, 2012.
† Meisner effect is the phenomenon of superconductor expulsion of magnetic field in the course of superconductor phase change into a superconducting state. Meisner and Robert discovered the phenomenon in 1933 when they measured the magnetic field outside superconducting tin and lead samples.

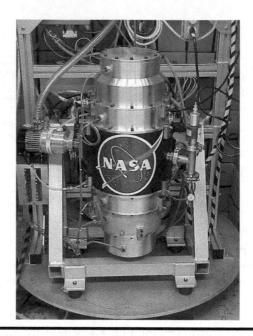

Figure 12.7   Flywheel module. (Photo by NASA.)

Figure 12.8   The superconductor in Meisner effect has extremely great potential for industrial applications. (Photo by Mai-Linh Doan.)

systems at present. Because transmission lines have resistance whose heating is in direct proportion to the square of current, there is great heat loss—it is calculated that in the present conventional transmission methods, heat loss results in waste of one-third of energy. Compared with common cables, superconducting cables have greater transmission capacity and smaller energy loss, so the voltage of the grid can be greatly reduced. For densely populated big cities, when power supply voltage remains unchanged, meeting the rapidly increasing power demand by expanding power transmission capacity within limited space, using superconducting cables instead of conventional copper wire cables, would be an ideal solution. With technological development and continuous development of new superconducting materials, superconductor power transmission will soon be realized, which will greatly transform the world's power application. Superconducting technology can also be used for making superconducting magnetic levitation trains that run without friction, greatly enhancing their speed and silence and effectively reducing mechanical wear. Using superconducting levitation to make bearings without wear can increase the rotating speed of bearings up to over 100,000 RPM.

Ultrahigh voltage. Power transmission is divided into high-voltage power transmission, extra-high power transmission, and ultrahigh power transmission based on different voltage class. Internationally, high voltage (HV) usually refers to 35–220 kV voltage; extra high voltage (EHV) refers to 330 kV and above, AC below 1000 kV, and DC below 800 kV voltage; ultrahigh voltage (UHV) refers to AC 1000 kV and above and DC above 800 kV voltage. The UHV grid is characterized by large transmission capacity, long distance power transmission, low line loss, less land occupied, and strong grid-connecting ability. But there is still controversy about UHV, especially regarding the economic efficiency and safety of AC UHV power transmission in the world. Since the 1970s, Europe, America, Japan, and other countries and regions all have carried out AC UHV power transmission technology research and experiments. Finally, only Soviet Union and Japan constructed AC UHV lines, but they failed to commercialize the technology. In recent years, China has become the leader in UHV technology and UHV engineering construction; it is the only country at present that has a commercially operated UHV grid.

Wireless transmission. It can be roughly classified into three kinds based on its power transmission principle. The first kind uses the electromagnetic induction principle. Nontouch type charging technology uses the principle of putting two coils in positions near each other. When current flows in a coil, the magnetic flux produced by it becomes the medium that leads to the other coil producing electrodynamic force. The technology is increasingly used in portable terminals. The second kind uses the principle with which antennas send and receive electromagnetic wave energy. This is basically the same as the radio principle that was used 100 years ago. The AC wave forms of electric waves are converted into DC in rectification circuits and used, but amplification circuits are not used. The technology has improved efficiency compared to the earlier technology. And manufacturers are putting it into use. The third kind uses the resonance of electromagnetic fields. Resonance technology is widely used in the electronic field. But in the power supply field, what is used is not electromagnetic wave or current but electric field or magnetic field. In November 2006, the research team of Marin Soljacic, assistant professor of the Department of Physics at Massachusetts Institute of Technology, first declared the possibility of using electric field in power supply technology.

Smart grid (see Figure 12.9). It is the new and modern grid formed by combining advanced sensing measurement technology, information communication technology, analysis and decision-making technology, automatic control technology, and energy and power technology and integrating them with the grid infrastructure. It includes power generation, transmission, transformation, distribution, use, and control. Based on an integrated high-speed two-way communication network, one can achieve the goal of a reliable, safe, economic, high-efficiency, environment-friendly grid through comprehensive use of different kinds of advanced technologies. Its main features include self-correction, driving and resisting attacks, providing the electricity that satisfies users' demand, allowing access to all kinds of power generation methods, starting power market, and optimizing high-efficiency operation. It has the advantage of being economic, highly efficient, clean, environmentally friendly, interactive, transparent, and open. It can work more flexibly, power dispatch effectively, and change users' behavior of electricity use by employing real-time electricity use information provided by advanced electronic ammeters to save

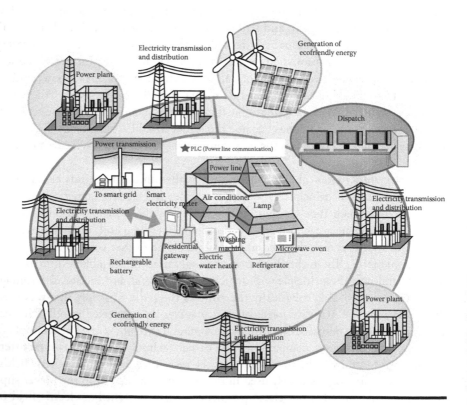

Figure 12.9   Diagram of a smart grid.

electricity. Different electricity prices can be fixed to further reduce peak use to avoid huge investment of adding power plants, which is very important for power suppliers and consumers and helpful for energy saving and emission reduction. It is more convenient for connecting wind power, solar power, and other kinds of energies that are unstable and easily affected by climate to the grid. With obvious advantages, smart grid represents the future of grid and becomes a key energy technology that each country will compete to develop.

The Rise and Development of Smart Cities

A smart city means an information city that enables sharing of information between the urban information infrastructure and systems and business operations, with the support of Internet of things, cloud computing, broadband networks, and other information communication technologies and through information perception, dissemination, and use, leading to improvement of citizens' living standards, urban operation and management efficiency, and public service levels; enhancement of economic quality and industrial competitive ability; and achievement of scientific and sustainable development.

The smart city plays an important role in solving the problems facing urban development; promoting industrial upgrading, economic structural adjustment, and social harmony; and raising national competitiveness. Many countries and government organizations have made plans of changing urban future development blueprints relying on Internet and information technology.

IBM announced in September 2009 that it would build America's first smart city at Dubuque, IA. The city will be "armed" by adopting a series of IBM's new technologies to realize complete digitization, connecting all urban resources to detect, analyze, and integrate all data, and intelligently respond to serve citizens' demands. Smart city construction in Sweden has a unique feature—transportation. Stockholm set up 18 roadside control stations along the roads leading to downtown areas to automatically identify the vehicle plates so as to collect a "road congestion tax" on the registered vehicles entering and exiting the downtown area during 6:30–18:30 from Monday to Friday in an attempt to reduce vehicle flow. It made traffic congestion fall 25%, traffic-lining-up time fall 50%, traffic exhaust emissions reduce 8%–14%, and greenhouse gas emissions fall 40%. The city of Inchon in South Korea announced that, in collaboration with Cisco, it will improve urban management efficiency based on network in an all-round way and try to forge a green, information based, convenient, seamlessly connected, ecological, smart city. Singapore launched the "Smart Country 2015" plan to make it a first-class international city with regard to economic and social development through information technologies including the Internet of things.

It is mentioned in the report of the Eighteenth National Congress of the Communist Party of China that to realize the synchronous development of industrialization, and informatization based urbanization with smart cities as representatives, and agricultural modernization is the driving force of the development of China in the next few years and one of the important approaches stimulating domestic consumption. At present, in China, smart city is rapidly being developed, and there are 28 above vice-provincial level cities proposing the goal of building smart cities.*

A smart city cannot happen without smart energy. China's urbanization path faces a severe situation of increasingly tight constraint of resources, serious environmental pollution, and ecological system deterioration; traditional energy cannot meet the demand of sustainable development. And the smart energy technologies represented by the aforementioned photovoltaic power generation, biomass power generation, compressed air energy storage, flywheel energy storage, superconducting transmission, wireless transmission, and other technologies will become the solution for easing the critical situation.

---

* *White Book of Intelligent City Technology*, China Institute of Communications and Intelligent City Forum, 2012.

Chapter 13

System Change: Paving the Way for Progress

13.1 Forward-Looking Development Plan

Adapting to outside changes is one of the key abilities that any country's energy supply system should have. Adaptation is a process that needs to use extraordinary wisdom to overcome the weakness of the energy system and to make a prospective energy development plan.

The energy development plan is a comprehensive arrangement for energy structure, development, production, conversion, use, and distribution made on the basis of the national economic and social development plan in a certain period and is also based on the associated energy demand prediction. It cannot be separated from the national economic and social development plan of a period. Energy is the important material base for national economic and social development, and the development of energy, society, and economy must maintain a reasonable and proportional relationship. Energy development is affected by national economic and social development; on the other hand, energy development also promotes or restrains national economic and social development.

Generally speaking, the energy development plan is made primarily with two goals: One is to guarantee energy safety and put forward a strategic goal and related supportive measures; the other is to press ahead sustainable development, a subject that receives the close attention of all the countries in the world. Focusing on the two goals, countries predict the future through advanced methods and technologies; most importantly, they predict future energy supply and demand. First of all, starting from the development plan of the macronational economy, they consider all available energy-saving measures and calculate the energy demands, adapting them to national economic and social development in accordance with scientific analysis and prediction. Second, proven energy reserves and existing production ability serve as a basis to make a reasonable plan of energy development and predict future energy supply. Third, the scale, structure, and regional distribution of energy demand and supply are considered; the abilities and all kinds of plans related to predicted energy demand and supply are comprehensively analyzed; and the quantity of capital, materials, manpower, and other demands needed for the plans are calculated. Finally, the energy plan with the best economic benefit and the most suitable for national economic and social development is made through a trial and error procedure of the aforementioned process.

The main contents of the energy plan include: Investigation and analysis of the present situation of energy supply and demand; energy demand prediction, including demand quantity and structure (department structure, space structure, and variety structure) prediction; energy supply design plan, assessment, and optimization; and plan test and decision. By rightly understanding the relationship between energy and economy, energy and environment, the part and the whole, short term and long term, and demand and possibilities, the energy plan will guarantee that the quantity, structure, and construction of all kinds of energy resources match with the national economic and social development, taking all factors into consideration and being comprehensive and reasonable.

Blueprint for a Secure Energy Future

On March 30, 2011, the U.S. government issued the *Blueprint for a Secure Energy Future*, which comprehensively depicted the national energy policy. On the same day, U.S. President Obama gave a speech at Georgetown University in Washington, DC, presenting the specific measures to be taken to realize energy goals and require one-third reduction of U.S. import of petroleum before 2025.

The *Blueprint for a Secure Energy Future* puts forward three strategies for future U.S. energy supply and safety: To develop and guarantee the U.S. energy supply, to provide consumers with alternatives of lower cost and better energy savings, and to realize future clean energy using creative methods.

While considering how to raise energy efficiency, reducing waste is an important problem that has to be solved. In America, residential and commercial buildings consume 40% of the total energy. The U.S. government has proposed new projects to help the U.S. people use new types of energy-saving building material, such as new lamps and lanterns, new windows, and new heating and cooling systems to update resident and commercial building. Obama said, "The good news in energy efficiency is that we have owned such technologies. What we need is to adopt encouraging measures to help merchants and users to install and use these energy-saving materials."

Blueprint for a Secure Energy Future requires us to "Innovate our Way to a Clean Energy Future: Leading the world in clean energy is critical to strengthening the American economy and winning the future. We can get there by creating markets for innovative clean technologies that are ready to deploy, and by funding cutting-edge research to produce the next generation of technologies. And as new, better, and more efficient technologies hit the market, the Federal government needs to put words into action and lead by example."

13.2 Promotion of High-Efficiency Industrial Organizations

Industrial organization means the organizing form and market relationship of all the enterprises in the same industry, including enterprise classification, structure, scale, distribution, monopoly, and competition in the market. It is the organizational carrier that helps enterprises carry out resource allocation in the course of growth, value creation, and realization, through abiding by industrial growth law and value law. Good industrial organization is vital to the long-term development of the energy industry, especially for the new type of energy industry that is being rapidly developed and relates to the future of energy.

Generally speaking, industrial organization tends to have the duality of developing scale economy and maintaining competitive vitality. It is generally believed that big enterprises can promote innovation because they have the advantages of funding, talent, and risk diversification. Therefore, the industrial organization that is most helpful for innovation is one that has a combination of appropriate monopoly and competition. To impel energy industry organization to be more efficient, the governments of all countries proactively adjust the relations of market exchange, competition and monopoly, market occupation, resources occupation, as well as energy industry structure, layout, and scale; they also regulate the relationship between enterprises, social organizations, and consumers to let them always be benign, cooperative, competitive, and efficient.

To raise the operation efficiency and management level of the energy industry organization and reduce operation cost, a lot of countries are loosening control of industries and carrying out structural reform; most of them now rely on commercialization, privatization, recombination, and introduction of a competitive system represented by a universal highly state-owned monopolized power industry. The state-owned enterprise into which commercialized management and operation pattern are being introduced is endowed with a certain pricing right, but it is separately taxed on its power generation, transmission, and distribution in accordance with an independent accounting method, by which it assumes its own losses and profits. Privatization refers to privatizing power companies or allowing private investment in power generation, transmission, and distribution. It should be kept in mind, though, that a privatized power company can also obtain monopoly privilege and present obstacles for the efficiency of the industrial organization. Recombination means vertically splitting the power department into power generation, power transmission, power distribution, and power sales companies with an independent legal person in accordance with different functions. In general, there is a natural monopoly in the power transmission and distribution field. A competitive system can be introduced in power generation and also a sales field to set up power generation and a sales market. There are many competitive patterns of power retail. One pattern allows many power generation enterprises to have their own distribution network and directly sell power locally. Another pattern lets power sales companies be independent but does not allow them to have power-generating equipment. The power sales companies buy power from power generating enterprises and resell it to end users; they have the functions of power distribution and sales at the same time.*

Each state is still enacting and improving laws related to guaranteeing energy industry development, making policies to encourage it, and establishing a socialized associated system. Social organizations of intermediary service and industrial organizations in the energy industry, such as industrial associations, enterprise alliances, mass organizations, and consumer groups, must be proactive. Industrial layout, scale, and structure are adjusted to give full play to the industrial agglomeration effect and scale effect, such as adjusting industrial space layout to facilitate industrial mergers and reconstruction or setting up comprehensive industrial parks, that is, ecological industrial parks. Establishing a technological research and development system for the energy industry helps to provide technological support for industrial development. For instance, to promote the development of the renewable resources industry, America invests a great amount of capital in industrial technical research and development and pays attention to adjusting national policies and industrial strategy according to domestic and overseas industrial development in order to guarantee the development and vigor of industrial organizations.

Global Wave of Market-Oriented Power Reform

The power industry has networked public, technical, and economical sectors. In the past, it was always regarded as a natural monopoly industry, and thus adopted vertical integrated monopoly operation. At the beginning of the 1980s, the slowing

---

* See *Finance of Energy*, Lin Boqiang and Huang Guangxiao, Tsinghua University Press, 2011.

down of economic growth in developed countries resulted in energy investment reduction, and developing countries' economic development promoted the rapid increase of power demand. It is the continuous progress made in power science and technology and the gradual increase of the knowledge of the power industry that stimulate energy investment and resolve the problem of system efficiency. Most countries in the world have launched a market-oriented reform of power. Great changes have been made in each country's power system through 30 years of reform and exploration.

Britain: Britain is the forerunner of market-oriented power reform. It issued the white paper Power Market Privatization in February 1998, marking the inception of the market-oriented power reform. The core of this reform lies in privatization and introduction of competition in the power market. The electricity price of resident users dropped 28% because of the introduction of competition and the rise of the ratio of low-cost natural gas power generation from 1% up to 22%. The electricity price of middle-sized industrial users dropped by about 31%. When many countries followed Britain's approach, Britain steamed ahead and came out with new power transaction rules to promote the integration of power generation and power sale and allow mutual mergers between power supply companies to realize scale benefit.

France: France opposes fragmentation to realize scale benefit. It formulated and carried out the Law of Power Public Service Development and Innovation under EU directives. The law mainly deals with the following: To make clear the mission and capital source of the public services; to set up a public service fund; to determine the opening time of the power supply market and provide users with the right of choice; to establish a production license system; to set up a power supervision committee, whose cost is covered by the government budget; and to allow provision of heating, gas, and other operating services to users with the right of choice. France proposed longitudinal integration to scale up the economy. It opposed fragmentation. It did not split the France Power Company but separated its functions of power generation, transmission, and distribution from finance.

EU (European Union): The EU has made it a point to establish a uniform power market. It passed the instructions of relaxing the power market in December 1996: Users with right of choice could freely choose power suppliers, and the 13 countries joining EU and EEA had to open their power supply market in accordance with a timetable; the power supply market could adopt third-party access, single buyer, and other business modes; for the power generation market, European countries could choose to adopt a bidding mechanism or license system to supervise new power generation capacity. By 2000, 80% of the European market had been opened up, far exceeding the prescribed 30%. European guidelines made many companies distinguish power generation, transmission, and distribution services and established different legal entities. It also eliminated European internal trade barriers. Competition promoted a price fall.

U.S.: America had two plans, and two results. The core of the U.S. power reform was to deregulate, and introduce, competition, improve efficiency, and lower power price. In 1992, the U.S. enacted the Energy Policy Act; it agreed to open the power transmission field and introduce competition in power wholesale marketing. In 1996, the Federal Energy Regulatory Commission planned to introduce an open power wholesale market; power plants and grid had to carry out functional separation and independent accounting. Just as American power reform was starting from California, some problems emerged in the course of relaxing power regulation because America believed that the market could settle everything, such as blackout, electricity price rocketing, and power companies applying for bankruptcy protection. But America's biggest eastern PJM grid succeeded by choosing the power reform mode of longitudinal integration in accordance with the real situation.

China: China broke through monopoly to promote competition. The original state power company controlled most of the capital of power generation and transmission before 2002. The company was split and recombined after the power system reform, and its power generation capital was directly reorganized or recombined into five nationwide independent power generation companies of roughly the same size. The China State Grid Corporation and China Southern Grid Corporation were established, and the State Electricity Regulatory Commission was set up (merged with the National Energy Bureau in March 2013, and affiliated to the National Development and Reform Commission). Although the reform still needs to intensify, the achievements are remarkable. The competitive pattern in the power generation side has primarily been formed; the internal constraint mechanism has been improved greatly; the vitality of power enterprises has obviously been strengthened; and the power supply ability has been greatly enhanced. The reform has not only maintained the safe stable operation of the grid but has also profoundly changed the Chinese power industry.

13.3 Continuous Improvement of Laws, Regulations, and Policies

Thoughtful energy laws, regulations, and policies are the most powerful support for guaranteeing energy safety. Every country in the world takes great effort to set up and improve related systems; encourage mechanisms in finance, investment, price, taxation, etc.; and pave the way for the rapid development of new energy.

Japan has grown into a power of new energy from a country poor in resources. This could not have been achieved without a sound legal foundation. The country enacted the Law of Special Measures for Promoting New Energy Use in 1997 with the purpose of promoting new energy and renewable energy, such as wind power, solar power, geothermal power, garbage power generation, and fuel cell power generation. The law was continuously improved in 1999, 2001, and 2002. China promulgated and implemented the Law on Renewable Energy in 2006 and the Law on Energy Saving in 2008 with the purpose of promoting energy saving, improving energy use efficiency, and protecting and improving the environment.

The setting up of new energy systems belongs to the category of national energy safety and is the state's and government's responsibility. Although laws, regulations, and policies provide strong support for energy system change, energy development and use is a capital-intensive technology; especially at the beginning, a great amount of capital is needed and investment risk is great. It is necessary for the government to enact economic support policies to guarantee and support change in energy systems. Governmental support is still not underestimated even in the fields where technology and market is mature.

Finance encouragement mechanism. The Japanese government allocates over 57 billion yen each year to ensure the smooth implementation of new solar plans, 63.5% of which is used for the development of new energy technology. By 2006, Japanese solar technology had been in use for a long time and was highly marketed with governmental support. Hence, the Japanese government canceled the subsidy toward the solar industry. Later, the Japanese domestic solar market fell into stagnation, the solar cell producers' initiative of investing in scientific research projects sharply declined, and the Japanese Sharp Corporation's status of being the first supplier of solar cells was replaced by German Q-Cell Corp. Facing facts, Japan had to restart its solar energy subsidy policy in 2009.

Investment encouragement mechanism. Its main purpose is to reduce the capital needed to spread new technologies and decrease investors' risk while promoting investment. To encourage a photovoltaic power generation system, Japan promulgated a capital donation plan in 1994. It not only effectively promoted the development of a photovoltaic power generation system but also provided support for the development of the related technologies of hydropower, geothermal power, and other new energies. The U.S. Department of Energy provided a loan guarantee of $500 million for geothermal energy use projects, promoting private enterprises to develop and use geothermal energy. Spain used preferential loan agreement, and banks guaranteed cash flow for related plans to encourage and support the development of wind power industry.

Price encouragement mechanism. Compulsory buying system is the most common price encouragement mechanism and has been adopted by many countries. In 1978, the U.S. required the power supply bureau to buy the power generated by small-sized power plants and combined heat and power generation plants that conformed to "qualified facilities" condition and also asked it to pay "saved cost," which meant increased cost for the power generated or bought through other approaches. The mechanism had obvious effects. By the 1980s–1990s, the renewable energy projects developed by the U.S. had exceeded 12,000 MW.

Taxation encouragement mechanism. Preferential taxation can lower the cost of new energy development and promote investment decisions. In 2008, the Malaysian government encouraged companies to install solar power generating devices in its budget; double preference of taxation reduction would be given to companies that had installed these devices. Taxation measures can still be used to settle the energy import dependency problem, environment pollution problem, and other external problems related to energy production and consumption. In the 1990s, Holland and Germany levied an environmental protection tax to energy end users, but Holland did not collect an energy tax on renewable an energy electricity consumption and established an environment protection fund using the tax revenue paid by non-renewable energy electricity consumers to encourage renewable energy development.

U.S. Policy of Supporting New Energy

In the face of an increasingly severe energy situation and ecological and environmental problems, the U.S. has gradually recognized the unsustainability of traveling the traditional industrialization road and mostly uses taxation policy, financial subsidy policy, and quotation policy to greatly support and develop new energy.

Investment subsidy. The U.S. practiced investment subsidy policy for wind power projects in the early 1980s. The total investment subsidy of the federal and state governments accounted for over 50% of the total investment at that time. The energy fund of the U.S. Department of Energy, the renewable energy fund of the Department of Treasury, and the rural energy fund of the Department of Agriculture were specially established to support new energy development.

Quotation system. The quotation system of renewable energy is a mandatory policy for power companies. It stipulates that power companies must provide users with the least proportion or quantity of renewable energy electricity and will be punished if they cannot meet the requirement. At the same time, they need to obtain a related green certificate when they sell renewable energy electricity. The certificate can be sold on special green certificate trade markets, and its price is determined by market supply and demand. This provides additional income for renewable energy power producers and promotes the development of renewable energy.

Taxation policy. U.S. taxation policy of renewable energy is a well-thought out scheme. It plays a very good role in stimulating enterprises, families, and individuals to use more energy-saving and clean energy products. The U.S. Energy Tax Law of 1978 stipulated that 30% of the investment of home owners who bought solar and wind power generating equipment could be deducted from the income tax paid in the same year and that 25% of the total investment of solar, wind and geothermal power generation can be deducted from federal income tax in the same year.

In 1992, the U.S. enacted the policies of deductible and subsidiary production in the Energy Policy Act. Deductible production meant that enterprises of wind power generation and biomass energy power generation could enjoy an exemption of 1.9 cents from individual or enterprise income tax each time they produced 1 kilowatt of electricity from the day they started production. Subsidiary production meant that a subsidy of 1.5 cents was given to duty-free public institutions, local governments, and countryside-run renewable energy power-generating enterprises each time they produced 1 kW of electricity. Besides, the Act still stipulated that enterprises that invested in solar and geothermal power could permanently enjoy 10% tax deduction preference.

In 2005, America promulgated the Energy Tax Act again, of which some measures played a promoting role in renewable energy use, such as encouraging the public to use solar energy. The act stipulated that house owners who installed solar water heaters could enjoy 30% tax reduction at most, and the U.S. federal government still allocated $1.3 billion to encourage private residence to use solar energy, which has zero pollution. Besides, consumers who bought new fuel cars can obtain a tax reduction of $34,000 at most. The production of highly efficient cars and energy-saving household electrical appliances would receive governmental tax preferences. The new act required that by 2011 and 2012, the amount of alcohol used as gasoline additive in America would increase to 7.5 billion gallons each year, an increase over onefold the present level.

In 2007, the U.S. promulgated the Energy Promotion and Investment Act, which proposed to prolong the time limit of production tax deduction of renewable energy, such as wind and allowing solar energy, to 2013. It also suggested tax deduction of billions of dollars for clean coal projects and carbon dioxide storage and greatly increasing the production of hybrid vehicles and biofuel.

There was a plan to include a tax deduction of $18 billion dollars in the financial bailout plan of $700 billion dollars signed by the U.S. president George W. Bush in 2008. Its main contents proposed: to extend wind power producers' production tax credit by 1 year; to increase the investment tax credit of household and commercial solar devices to 30% within 8 years, which would cost $2.5 billion; to present tax preference plans for improving individual and enterprise energy efficiency; and to provide new tax credit policies for enterprises that adopt advanced carbon sequestration technology.

Chapter 14

International Cooperation: Finding Win–Win Solutions

14.1 Crisis Brings Cooperation

Mankind generally suffer heavy loss when a crisis occurs; they then learn from the bitter experience and eliminate all previous obstacles to cooperation—nowadays, no country fights over small trifles for their own interests; instead, they try to reach a consensus through compromise. This is how international energy and environment cooperation first came about. The first and second petroleum crises inflicted heavy loss on the oil-consuming countries, which did what they thought was right. The global warming and environment damage caused by developing and using energy also let mankind recognize the approaching danger that no one country can cope with alone. To avoid a collapse in global economy and the environment, all countries have to join hands and cooperate to save energy and the environment.

International environmental cooperation is mainly promoted by the United Nations. In June 1972, the UN held the first ever mankind and environment conference in Stockholm, Sweden, and discussed and passed the famous Declaration on the Human Environment, which asserted that protection of the environment was the responsibility of all humanity; this made the environmental protection movement reach the governmental level. Ideas on sustainable development were formulated as people gained a deeper understanding of equity (intergenerational equity and equity across generations) as a social development goal and reached a consensus on a series of global environment problems. The UN established the World Commission on Environment and Development in November 1983. The commission officially put forward the mode of sustainable development in the report *Our Common Future* in 1987. The 21st Century Agenda and UN Framework Convention on Climate Change were passed at the UN Conference on Environment and Development held in 1992. The main goal of the convention was the control of man-made greenhouse gas emission, to make it not result in climate change, and quantization of global greenhouse gas emission reduction. A binding emission reduction agreement—the Kyoto Protocol—was reached in December 1997 to distribute respective emission reduction responsibility. It stipulated that industrialized countries should reduce greenhouse gas emission by 5% between 2008 and 2012 compared to 1990.

The heads of about 100 countries, including China, the U.S., Japan, France, and the representatives of the 192 member states attended the UN Climate Summit in September 2009 to discuss how to cope with the global climate change problem and promulgate measures to control greenhouse gas emission. Three months later, the contracting parties of the UN Framework Convention on Climate Change convened the World Climate Conference in Copenhagen, Denmark, to discuss the implementation of each country's emission plan in the Kyoto Protocol and formulate a subsequent plan.

The then UN Secretary-General Ban Ki-moon announced the initiative of "everybody enjoys sustainable energy" at the UN General Assembly on November 1, 2001. The initiative tries to realize three goals by 2030: To ensure global popularization of modern energy service, to double the speed of enhancing energy efficiency, and to double the proportion of renewable energy in global energy consumption. Ban Ki-moon said that energy played a decisive role in economic development, but fossil-fuel-based energy was the important factor leading to climate change and also greatly affected poor populations. The three goals will promote equity, stimulate global economy, and help to preserve the ecology. He emphasized that sustainable energy would benefit all people. The barriers to promote sustainable energy were political will and the lack of resources in public and private sectors. To mobilize the initiative of developed and developing countries and emerging economies, he announced the setting up of a "senior team to realize the initiative of everyone enjoying sustainable energy," designated over 30 leaders from commerce, finance, government, and Non-Governmental Organizations as the team members to implement the specific action program of realizing the initiative goal and to start new sustainable energy cooperation.

At present, there are two main modes in international energy cooperation. One mode is multilateral cooperation with international organizations as a platform—this includes the cooperative organization of energy-importing countries, such as the International Energy Agency; the cooperative organization of energy-exporting countries, such as the Organization of Petroleum-Exporting Countries; the dialogue and cooperative mechanism between importing countries and exporting countries, such as the International Energy Conference, World Energy Council, World Petroleum Congress, and World Energy Chart Organization; the cooperation at the forums of international organizations or in dialogue framework, such as Group of Eight, United Nations Conference on Trade and Development, Asia-Pacific Economic Cooperation, and Association of Southeast Asian Nations. These international organizations all internally set up a working group on energy cooperation. The other mode is bilateral cooperation between governments, including the internal and mutual cooperation of energy-exporting countries, energy-importing countries, and transit countries. Multilateral cooperation between governments is relatively less compared to international organizations.

At present, the main international cooperative organizations in energy are the following.

IEA (International Energy Agency). IEA is an energy organization between governments consisting of 27 member states, affiliated with the Organization for Economic Cooperation and Development (OECD),* and is situated in Paris, France. It is dedicated to preventing unusual changes in petroleum supply while providing statistics of the international petroleum market and other energy fields. During 1973–1974, a relatively serious petroleum supply crisis broke out among the main western petroleum-importing countries because of the Middle East War, and crude oil price surged from less than $3 to over $13 per barrel. This had a great effect on each country's economic development, forcing them to look for a solution.

On November 18, 1974, OECD organized 16 of its then 24 countries to hold a working conference, and they signed the Agreement on International Energy Plan and established the IEA.†

Organization of Petroleum Exporting Countries (OPEC). On September 14, 1960, Iran, Iraq, Kuwait, Saudi Arabia, and Venezuela formed OPEC to coordinate the petroleum policies of member states, to determine crude oil production and price through joint negotiation, and to take concerted action against western countries' exploitation and pillage of petroleum-producing countries to protect their own resources and interests. OPEC's headquarters was situated in Geneva, Switzerland, but moved to Vienna, Austria, later. In addition to the original five countries, OPEC now includes Algeria, Libya, Nigeria, Qatar, The United Arab Emirates, Angola, and Ecuador and has developed into an international petroleum organization covering the main petroleum-producing countries in Asia, Africa, and Latin America, accounting for over 78% of the total petroleum reserves in the world, 40% of global petroleum production, and 50% of global petroleum export. OPEC petroleum production stability and decisions have important influence on international petroleum price.

---

* Organization for Economic Cooperation and Development (OECD) is an intergovernmental international economic organization consisting of over 30 countries with a market economy. Its purpose is to work together to cope with the challenges in economy, society, and government governance brought about by globalization and to seize the opportunities brought by it.
† See China Energy Development Report (2009), Zhang Guobao, Economic Science Press, 2009.

Energy Charter Organization. It is an international energy organization with the Energy Charter Treaty (with the European Energy Charter as its predecessor) as its core. The Dutch ex-prime minister Rudd Lubbers suggested the establishment of a European energy community and a European energy charter at the summit held in Berlin, Germany, in June 1990, which was attended by the heads of the 12 EC nations. The representatives of all the states of the EU signed the declaration—the European Energy Charter—in December 1991.*

International Renewable Energy Agency (IRENA). It was established under the background of the increasing seriousness of the global warming problem, fossil fuel depletion, and increasing expansion of the international interest in renewable energy. Its main functions are to provide policy consultation on renewable energy to member states, promote the technical transfer of renewable energy, support constructing the industry service system of renewable energy and other institutions, carry out the talent training needed to generate renewable energy, set up an information platform and industrial standards of renewable energy, and form an international investment and financing mechanism of renewable energy. The German government proposed to establish the agency for the first time at the First International Renewable Energy Conference held in 2004. It started work in 2008 and presented and discussed agency tenets, decision mechanisms, capital plans, and other regulation drafts. In January 2009, IRENA held the founding conference at Bonn, Germany; 75 countries, including Germany, Denmark, and Spain, the three founding nations, became full members; China, the U.S., Japan, and other countries attended the conference as observers.†

World Energy Council (WEC). It is a comprehensive international energy non-government academic organization. Its purpose and task are to proactively research and help all countries solve energy problems; promote the sustainable development and use of world energy under the condition of it being beneficial to all countries; research the problems of production, transportation, and usage methods of potential energy and other kinds of energies; explore the relationship between energy consumption and economic increase; and collect and communicate data of energy or resources use. It was set up on July 11, 1924, in London and was originally called the World Power Conference. Twenty-four countries participated. It was paused during the Second World War, restarted in 1950, changed its name to the World Energy Congress in 1968, and finally changed to its present name in January 1990.

Intergovernmental Panel on Climate Change (IPCC). It is an intergovernmental organization set up by WMO (World Meteorological Organization) and UNEP (United Nations Environment Program) in 1988. Its secretariat is situated within the WMO in Geneva, Switzerland. Its main task is to assess the scientific recognition of climate change and possible measures of adapting and slowing down climate

---

* See *Global Energy Trend*, Center for Economic Security Studies of China Institute of Contemporary International Relations, Current Affairs Press, 2005.

† Edited by Feng Qi, published by Xinhua News Agency, October 4, 2006.

change. The conclusion of the IPCC assessment is an important scientific basis on which the international community commonly takes action of coping with climate change. It has a decisive position in foreign negotiations on climate change and international action and plays an important role through the UN Framework Convention on Climate Change and the Kyoto Protocol. At present, IPCC has made four assessments altogether, namely 1990 (FAR), 1995 (SAR), 2001 (TAR), and 2007 (AR4). The fifth assessment report (AR5) will be finished in 2014. The conclusion of IPCC assessment reports on the possibility of global warming caused by human activities is grim and convincing.

UN Framework Convention on Climate Change

UN Framework Convention on Climate Change (UNFCCC) is the convention reached by the UN Intergovernmental Negotiating Commission on climate change on May 22, 1992, and passed at UNCED (United Nations Conference on Environment and Development) held in Rio de Janeiro, Brazil, on June 4, 1992. It is the first international convention held solely to discuss controlling carbon dioxide and other greenhouse gas emissions to cope with the global warming that is having an adverse effect on human economy and society and forms a basic framework by which the international community can cooperate in coping with global climate change.

Its second article stipulates that the final goal of any related legal "document passed by this convention and its parties is the stabilization of greenhouse gas concentrations in the atmosphere at a level that would prevent dangerous anthropogenic interference with the climate system. Such a level should be achieved within a time-frame sufficient to allow ecosystems to adapt naturally to climate change, to ensure that food production is not threatened and to enable economic development to proceed in a sustainable manner."

The members of the convention have convened Conferences of the Parties (COP) each year since 1995 to assess the progress of coping with climate change. The Kyoto Protocol was reached in 1997, making greenhouse gas emission reduction a legal obligation of developed countries.

The Kyoto Protocol stipulates that the emission of carbon dioxide and five other kinds of greenhouse gases in all developed countries should be reduced to 5.2% of that seen in 1990, and this should be done by 2010. The specific goals of the emission reduction that all developed countries must accomplish during the period from 2008 to 2012 are that compared with 1990, the European Union will reduce its emission by 8%, United States by 7%, Japan by 6%, Canada by 6%, and eastern European states by 5%–8%; however, New Zealand, Russia, and Ukraine will maintain the same level of emission as in 1990. At the same time, it also allows Ireland, Australia, and Norway, respectively, to increase their emission by 10%, by 8%, and by 1% compared to that of 1990. The UNFCCC 18th Conference of the Parties and 8th Conference of the Parties was held in Doha, Qatar, from November 26 to December 7, 2012. The conferences finally reached a consensus on the second commitment period of the Kyoto Protocol that will be performed starting from 2013. A consensus was reached on the second commitment period with a period of 8 years. They also passed many resolutions in long-term climate capital, UNFCCC long-term cooperation working group results, Durban platform, and loss and damage compensation. However, Canada, Japan, New Zealand, and Russia did not participate in the second commitment period of the Kyoto Protocol.

14.2 Many Hands Make Light Work

The egocentrism of western countries and the following of their own agenda in the three petroleum crises made them suffer great loss and taught future generations a lesson: In the energy sector, the interest of all participants can only be guaranteed when all countries make concerted efforts for the same purpose. After the International Energy Agency was established, massive international energy cooperation gradually began. The Organization of Petroleum Exporting Countries, Independent Petroleum Exporting Countries Group, Group of Eight, International Energy Congress, World Energy Council, European Union, Asia-Pacific Economic Cooperation, Association of Southeast Asian Nations, Shanghai Cooperation Organization, and other groups and organizations also carried out the tenet of international energy cooperation and protected energy safety by gathering the advantages and abilities of all parties so as to achieve mutual benefit (a win–win situation) and common goals. The international energy cooperation has embodied higher and higher values through unremitting efforts.

Promote reasonable energy development and guarantee energy safety and stability: Take China as an example. Sustainable development has been an important issue for a long time, but China's energy consumption has been increasing, especially the consumption of petroleum energy, which has brought a great threat to its energy safety. It is very necessary to build a "special" cooperative relationship with energy-producing countries through bilateral cooperation to some extent to raise the safety coefficient of the Chinese energy supply and guarantee the stable supply of strategic resources. The Chinese government began to support energy enterprises early to develop their market globally under "bilateral longitudinal cooperation." The China National Petroleum Corporation (CNPC) bought the shares of oil fields in Thailand, Canada, and Peru in 1993, bringing in the era of petroleum and gas cooperation and integrating it with the China National Offshore Oil Corporation (CNOOC) in Southeast Asia and West Asia. It obtained the right of development of Iran's biggest gas field, South Pars, and the biggest land oil field, South Azadegan, in 2009. China has participated in the development of over 70 oil and gas projects in the world, gained about 50 million tons of oil, accounting for about 25% of the total petroleum import. Natural gas supplies for China have also been diversified in Middle Asia, Russia, and Myanmar. (In order to keep a safe natural gas supply, China imports from these countries.) These cooperation projects have not only made local energy more reasonably developed but also effectively supplemented Chinese energy supply, guaranteeing energy safety and stability.

Promote energy technology innovation and improve energy efficiency: China adopted an extensive energy development pattern with backward technological equipment for a very long time, and mining efficiency was much lower than America and European developed countries. Specifically, the Chinese coal mining mechanization rate was 45%, far behind the global advanced level of 80%–100%; petroleum mining efficiency was only 20%, far lower than the international average level of 60%; energy efficiency has been low, energy consumption per unit output value was 2.3 times the world level. In contrast, EU countries have made huge investments in the development and use of new energy, and related industrial technologies have been greatly developed. They proposed new energy policies in 2007 and intensified the support for the new energy policies after the financial crisis of 2008. According to the estimation by EU's Department of Commerce, there is great potential in the Chinese clean energy market. China and the EU are complementary and have a very extensive cooperative space in low carbon new energy fields; developing new energy has become the common goal of China–EU energy cooperation. In the field of wind power, EU is the important technical supplier of wind power generation. It has world-class wind power generation enterprises, such as Siemens, Vestas, etc. They all have established branches in the form of sole proprietorship or joint ventures, providing a series of key technologies and talents training services. In biomass energy power generation, Danish BWE, a world-leading company, is also in technical cooperation with China, providing great technical support for the Chinese energy industry.

Lower global greenhouse gases emission and accelerate energy transformation: Energy is closely related to climate and global environment and has become an important challenge to human subsistence. The world energy structure is in oscillation, adjustment, and change. Energy that is intertwined with economy, trade, foreign affairs, and other problems has become more complex. Under such a background, it is especially important to strengthen environment protection, speed up energy transformation, and promote renewable energy development by cooperation. Take China and Denmark as an example: the two countries are vigorously developing and using renewable energy through bilateral cooperation at the government level. They have developed the China–Denmark wind power development project, the China–Denmark biomass energy clean development mechanism project, and the China–Denmark renewable energy development project since 2005. These projects have provided good economic and social benefit, promoted the two nations' energy transformation, made contributions to global renewable energy development and coping with greenhouse gases emission, climate change, and other problems, and have become an important model of bilateral cooperation.

Many Countries Work Together to Fill the Leak of the Ozone Layer

In the 1930s, the U.S. E.I. Du Pont Company synthesized a kind of material, named freon, a chlorfluorocarbon (CFC). It is widely used as a refrigerant and foaming agent because it is chemically stable, nonflammable, and nonpoisonous, and hence it was regarded as a significant invention then. However, its good reputation ended when people discovered in the 1970s that CFCs damaged the ozone sphere (see Figure 14.1).* The ozone sphere is like a protective umbrella of the earth that prevents animals and plants from being harmed by ultraviolet radiation. The damage to the ozone sphere not only does great harm to the human body but also affects the entire ecological balance, threatening the existence of life.

And then, the world acted. The governing council of the UNEP held an international conference to evaluate the ozone sphere in April 1976. An expert meeting was held in Washington, DC, in March 1977 and attended by 32 countries. "The world plan for ozone sphere action" proposal was passed for the first time. In 1981, the council set up a working group with the mission of preparing for a global convention to protect the ozone sphere. In 1982, a Japanese Antarctic Fleet observed and discovered the ozone sphere over the Antarctic obviously thinned out and formed an "ozone hole." Therefore, the appeal to protect the ozone sphere became internationally louder, and the working team sped up their progress.

The Vienna Convention for the Protection of the Ozone Layer was passed in Vienna, Austria, in March 1985 through 4 years of hard work and was put into effect in September

---

* Edited by Feng Qi, published by Xinhua News Agency, on October 4, 2006.

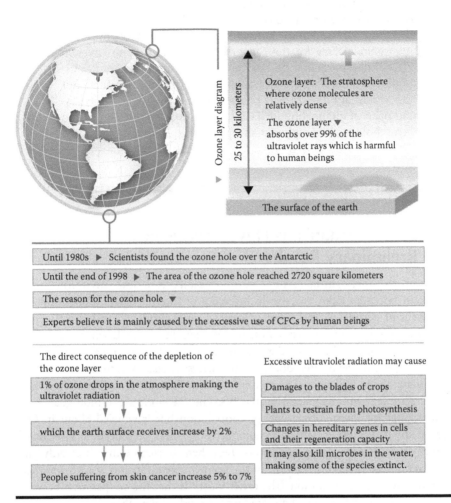

Ozone layer diagram

25 to 30 kilometers

Ozone layer: The stratosphere where ozone molecules are relatively dense

The ozone layer ▼ absorbs over 99% of the ultraviolet rays which is harmful to human beings

The surface of the earth

Until 1980s ▶ Scientists found the ozone hole over the Antarctic

Until the end of 1998 ▶ The area of the ozone hole reached 2720 square kilometers

The reason for the ozone hole ▼

Experts believe it is mainly caused by the excessive use of CFCs by human beings

The direct consequence of the depletion of the ozone layer

1% of ozone drops in the atmosphere making the ultraviolet radiation

▼ ▼ ▼

which the earth surface receives increase by 2%

▼ ▼ ▼

People suffering from skin cancer increase 5% to 7%

Excessive ultraviolet radiation may cause

Damages to the blades of crops

Plants to restrain from photosynthesis

Changes in hereditary genes in cells and their regeneration capacity

It may also kill microbes in the water, making some of the species extinct.

Figure 14.1   Serious ozone loss above Antarctica.

1988. Based on it, the Montreal Protocol on Ozone Depleting Substances was passed at the Montreal Conference, held in Canada, on September 16, 1987, and came into effect on January 1, 1989. The protocol stipulated that each member organization (state or state group) would freeze and reduce production and consumption of five kinds of freon and three kinds of bromides in accordance with the reduction timetable.

East Timor joined the Montreal Protocol on September 16, 2009, and made it the first protocol that was approved by all the sovereign states in UN history, fully showing the resolution of the whole world to protect the ozone layer. Ban Ki-moon, the then UN secretary general, gave a speech on the 17th International Day for the Preservation of the Ozone Layer in 2011, "all contracting parties have made an important contribution to protecting the global climate system for 24 years after the Montreal Protocol was passed and now agree to speed up the pace of eliminating HCFCs." It is estimated that the ozone layer hole will completely disappear by 2050 owing to the close cooperation and concerted efforts of all the states in the world.

14.3 Such Mechanisms Have Produced Preliminary Results

At present, the uncertain factors of the international economic environment, the increase in potential risks, the rise in intensity of international competition, the high prices of petroleum and other primary energy products, and the effects of some international political factors on world economy is far beyond the control scope of one country's ability. Under such international Geo-political environment, to make full use of the limited domestic and overseas resources to deepen international bilateral and multilateral energy cooperation in order to realize mutual benefit and achieve a win–win situation has become the world's common goal. Each country longs for international energy cooperation but is worried about it. One worry is divided interests. Today's cooperative partners may become tomorrow's enemies or rivals, attempting to oppress the other, so each country not only pays close attention to their contribution to the cooperative effort but also to who benefits more from the cooperation. The other worry is that the cooperation between countries lacks guarantee and therefore faces the risk of default.

Although these factors make energy cooperation between countries unstable, strengthening the cooperation is imperative. Much effort is needed for coordination to ensure common energy safety.

Increase information sharing between countries and set up a mutual trust mechanism. Although there are common interests in energy and an urge for cooperation between nations, diversified communication activities are still needed to realize information sharing and promote mutual understanding to build mutual trust, to form an effective mutual supervision channel, and to have a distribution pattern of cost and benefit. The mutual understanding of the basic energy state, policy inclination, cooperation, and goal can be increased through high-level forums, common studies, and information-sharing mechanism in a determined energy cooperation field.

Set up an international energy cooperation system: An international system is an important means for maintaining international energy cooperation. Each nation's worries can be eliminated through three measures to strengthen cooperative stability: form a common expectation and legalize action to raise "reputation" cost of breach; provide an information exchange platform to decide the transaction cost in cooperation to facilitate supervision over breach and increase difficulties of breach; and reinforce mutual benefit and win–win situation and institutionalize a mutual benefit system to punish violators more easily.

Shanghai Cooperation Organization Promotes Regional Energy Cooperation

In June 2006, at the Shanghai Cooperation Organization* summit held in Shanghai, China, Kazakhstan proposed to enact an "Asian energy strategy," and Russia suggested to establish an "energy club" in the framework of the Shanghai Cooperation Organization, the purpose of which was to coordinate member states' energy mining and transportation plans and make the club a model for considering the interests of both petroleum and natural gas exporting countries and importing countries.

On September 15, 2006, the Shanghai Cooperation Organization held a government head conference in Dushanbe, the capital of Tajikistan. The Shanghai Cooperation Organization Energy Working Team studied and established the energy club of the organization. The representatives of five countries unanimously agreed to set up the club on June 29, 2007, at the ministerial conference of energy held in Moscow, to principally reach a consensus on the basic regulations of the organization and to position the objective of the energy club to "form a regionally uniform energy space."

The energy cooperation of the Shanghai Cooperation Organization is divided into three steps: the first step is to set up a coordinating institution to promote member states' connection in energy and economy, including promoting the implementation of a Multilateral Economic and Trade Cooperation Program and other important documents of the organization; the second step is to conduct project cooperation, carry out modernization transformation of the existing energy industries of the member states through an established cooperation

---

* The members of Shanghai Cooperation Organization (SCO) are China, Russia, Kazakhstan, Kyrgyzstan, Tajikistan, and Uzbekistan.

platform, develop energy transportation infrastructure, prospect and develop new oil fields and gas fields together, create conditions for mutually entering power markets and transmitting, develop energy saving technologies together, and foster professionals in management and technology; the third step is to coordinate or unify energy policies and promote the legislation of economic policies, laws, and regulations under the prerequisite of successfully resolving strategic and technological problems, such as to lift the control over energy price, unify energy transportation costs, formulate a uniform taxation system, and eliminate malignant competition.[*]

[*] See *Middle East Energy and Superpower Games*, Zhang Ning, Changchun Publishing House, 2009.

Chapter 15

Conservatively Optimistic: There Are Still Disappointments among Hopes

15.1 Technical Bottlenecks

The rapid increase of energy consumption, energy shortage, climate and environment deterioration, and other problems requires better technologies in the energy sector. Although great progress has been made, whether they are technical innovations of traditional energy or technical exploration of new energy, they all cannot meet the real and potential requirements for energy; besides, they are beset with technical bottlenecks to different degrees.

Traditional energy technology. Over 100 years after the Industrial Revolution began, a big breakthrough has been made in traditional energy technologies, but it does not indicate that traditional energy technology is perfect. No matter how effective safety devices are, accidents inevitably take place again and again. Take power as an example: The electricity generated by power plants is transmitted to end users through the grid. Even if one link goes wrong, it can result in a serious accident similar to the European blackout in November 2006. There were two reasons for the accident: one was the European cold weather leading to increase of electricity demand; the other was the shutdown of an electric transmission line to let a transport ship pass across the river, which increased pressure on the normal operation of the whole grid.*

Solar technology. Solar energy is clean and nonpolluting, but the manufacture of its equipment is not completely green. For example, polycrystalline silicon, the main material of solar batteries, will produce the byproduct silicon tetrachloride during the course of production, which is very corrosive, severely irritates eyes and respiratory passages, and causes skin tissue necrosis after contact. If the recycling process in the course of production is not thorough, these harmful substances containing chlorine are extremely likely to leak and cause serious latent safety and pollution concerns. The surface glass of urban photovoltaic panels and solar water heaters reflect strong light under the sun, causing light pollution. The retina and iris of those who work and live for a long time in the environment of light pollution will be damaged to different degrees, resulting in loss of eyesight and 45% more chance of getting cataracts; they will also suffer from dizziness, dysphoria, insomnia, loss of appetite, low spirit, feebleness, and other neuroses.

Hydropower technology. Hydropower generation is criticized for upsetting the ecological equilibrium. To build a dam and generate power using river water means that the reservoir area of the dam is inevitably submerged, which results in loss of forest, grassland, and wild animal and plant habitats; reduction of species number and diversity; and environment degradation of the upstream watershed, which obstructs the passage of creatures, leads to the extinction of migratory fishes, and damages the diversity of aquatic animals and plants. The U.S. started the biggest dam-demolishing project in history in 2011—two water power generating dams with a history of over 100 years were dismantled to eliminate the obstruction of the migratory spawning of the salmon in Elwha River.

Wind power generating technology. Wind power generation affects the health of habitants nearby. Residents who live near the wind power plant in Shizuoka Prefecture, Japan, where Fujiyama is, suffer from shoulder rigidity, headaches, insomnia, hand tremors, etc. Their symptoms were alleviated whenever the turbines stopped because of mechanical failure or other reasons. The relation between these troubles and turbines is not clear, but the infrasonic waves produced by the turbines are very likely to be the cause. The waves vibrate 1–20 times/s and cannot be heard by human ears because their frequency is too low. At present, similar complaints have also been reported in other areas of Japan. Some experts assert that infrasonic wave noise harms human health. Serious problems will rise if measures are not immediately taken. Wind power generation is also a threat to birds—they hit the rotating blades of the turbines and fall to their death. The Japanese Ministry of Environment has verified that 13 rare white-tailed eagles have lost their lives because of this from 2003–2013. This is just the tip of the iceberg of the harm done by the turbines.

---

* See *Future of Energy*, Phil O'Keefe, et al., Petroleum Industry Press, 2011.

Hydrogen generation technology. Hydrogen supply is the biggest problem in hydrogen energy car development at present. The existing technology uses hydride as the hydrogen-storing material and utilizes waste heat of engine coolant and exhaust gas to release hydrogen. Although hydrogen hybrid cars have been successful on buses, mail cars, and cars, the cars with such technology are still not very mature at present. Hydrogen generation technologies still need a major breakthrough.

Flammable ice technology. It is difficult to eliminate the adverse effect of flammable ice mining on the environment. Flammable ice intensifies the greenhouse effect. Methane is an intense greenhouse gas second only to carbon dioxide, and the methane contained in global flammable ice is about 3000 times that of the atmosphere. The methane produced by flammable ice decomposition will obviously accelerate the course of global warming. Flammable ice damages the ocean environment. The methane entering sea water oxidizes relatively fast, affects its chemical property, and consumes a great amount of oxygen from sea water, thereby leading to an environment lacking oxygen. It harms marine microorganism growth and possibly results in sea water evaporation and seismic sea waves. When flammable ice is mined, it decomposes and releases great amounts of water, loosens rock strata, and causes disastrous geological changes like submarine slumps. Flammable ice can also lead to well drilling deformation and increase the risk to off shore oil rigs if a great amount of flammable ice decomposes in the course of drilling.

Carbon capture technology. This technology refers to the means by which carbon dioxide in the atmosphere is captured, compressed, and stored in depleted oil or natural gas fields or other safe underground places. The technology can effectively reduce the greenhouse gases produced by combustion of fossil fuel. It also seems easy. Carbon dioxide and amine substances can react, combine at low temperature, and separate at high temperature. Therefore, the carbon dioxide in the waste gases of power plants can be separated by allowing the gas to pass through amine liquid and then released from amine liquid by heating it. But there are still technical and economic obstacles to extensively use carbon capture. First, the cost of carbon dioxide capture is extremely expensive. Research conducted by Massachu-setts Institute of Technology points out its cost is as high as about $30/ton. Second, the buried places must be tested for faults, or great amounts of greenhouse gas will likely be rereleased into the atmosphere if there is an earthquake or other geological change.

Fukushima Nuclear Power Plant Accident in Japan

A magnitude 9.0 earthquake took place off the eastern coast of Miyagi-ken, Japan, on March 11, 2011, and caused a seismic sea wave. It resulted in the damage to the equipment of the Fukushima Daiichi Nuclear Power Plant. The reactor core meltdown, radiation release, and related disasters of this plant has made it the most serious accident since the Chernobyl disaster of 1986 (see Figure 15.1).

There are six boiling water reactor units in the Fukushima Daiichi Nuclear Power Plant, which were developed and designed by General Electric Company and managed and operated by Tokyo Electric Power Company. When the earthquake took place, units No. 4, 5, and 6 were stopped for regular inspection, and units No. 1, 2, and 3 automatically stopped immediately upon detecting the earthquake. Hence, the plant's power generating function stopped. The connection between the units and the grid was massively damaged, and only diesel generators could be relied on to drive the electronic and cooling systems. But they were damaged by the coming seismic sea wave. The cooling system stopped because of this, and the reactors began to overheat. The reactor cores of No. 1, 2, and 3 had a meltdown over the next several hours to several days. Employees tried to cool the reactors, but had little success. During this time, several hydrogen explosion accidents happened.

Figure 15.1   The Fukushima I Nuclear Power Plant after the 2011 Tōhoku earthquake and tsunami. (Photo by Digital Globe.)

Chief Cabinet Secretary Yukio Edano issued emergency evacuation instructions on March 12, requiring the residents within 10 km around the plant to be immediately evacuated. Their number was about 45,000; later, the evacuation radius was expanded to 20 km.

The Japanese Nuclear and Industrial Safety Agency upgraded the accident to level 7—the highest level in the International Nuclear Event Scale on April 12. It is the second accident after the Chernobyl accident that has been rated as level 7. A great amount of radioactive substance was released into the land and sea during the accident. The Japanese government detected radioactive cesium with high concentration at areas 30–50 km away from the plant, which was extremely worrying. The Japanese government and Tokyo Electric Power Company were criticized by foreign media because they did not keep the public informed and failed to effectively manage the emergency.

15.2 Defects in System Structure

In addition to technology, the energy system is not perfect. To some extent, the energy system, whether involving domestic or international cooperation, is more complicated than energy technology because the energy system involves many stakeholders. It is a balancing act between energy interests, economic interests and even political interests. Reaching this balance takes a very long time. The 1993 Nobel Economics Prize winner Douglass North has a famous "path dependence" theory: In the course of social and economic system cooperation, once people make certain social choices, it is like taking a path of no return; inertia will make the choice a continuous self-reinforcement, it will be very hard to leave the path. According to the theory, the party whose interests are affected will not appreciate the new systematic change or innovation and even oppose it by hook or crook, so the effect of the energy system innovation will be greatly discounted.

Defects of domestic energy system. For a long time, the energy industry excessively relied on energy policy and an operation mechanism that was not completely reasonable; the management system was not very clear and related policies needed to be improved.

Let us consider the first problem of excessively relying on energy policy, particularly the subsidy of renewable energy. For example, European support for solar energy was the greatest. But after 2008, Spain and other countries slashed subsidies; Germany and Italy, as the first and second solar market, made plans to limit subsidies one after another. The solar market experienced turbulence. Even South Africa adjusted its wind power incentive policy. These all added uncertainties to the development prospects of the new energy industry.

It is taken for granted that developing countries and transforming countries have unreasonable operation mechanisms. For example, the Chinese energy price forming mechanism has not been reasonable for a long time, and the role of market in energy supply and demand is not fully exerted. China had carried out a low-price energy policy for a long time. When international energy prices, especially crude oil price, go up and energy-producing enterprises suffer losses, the government tends to subsidize them; when coal price rises and power enterprises suffer losses, their bills are also usually bought by the government. Although this mechanism helps to stabilize energy prices, because they cannot reflect the supply and demand relationship, it results in inefficient resource allocation and practically stimulates the excessive development of high energy-intensive industries. Thus, a pattern of excess capacity is formed in the end.

It is also common for developing countries to have unclear management systems. For example, the work of renewable energy and new energy in China is scattered across many departments, resulting in poor coordination of the departments at different levels and management chaos. The overcomplicated methods and procedures taken for developing renewable energy set excessive obstacles for project development and discourage developers and investors from entering the market.[*]

It is also obvious that the energy policy needs to be improved, particularly resource taxes and fees. For example, the main problems in the Chinese resource tax are: specific tax and quota tax cause slow income increase; resource tax and fee relationship is confusing, its levying is nonstandard; the distribution of resources benefit is not reasonable. Lagging resource tax and fee policies and excessive mining of resources cause serious ecological and environmental pollution problems.

Defects of international energy cooperation system. The international energy cooperation system, jointly established and maintained by many international energy cooperation organizations, plays a positive role in energy safety, production, consumption, and pricing. However, its limitation is also relatively obvious.

The role of IEA is important, but it is difficult to hear its voice. The functions of OPEC are only limited to providing services to petroleum-producing countries. The dialogues between IEA and OPEC to promote petroleum market transparency still continue, but without substantial progress so far.

The World Energy Charter Organization also has no substantial influence on the energy market, though it tries to integrate the eastern and western European energy systems. Russia, an important supplier to Europe, believes that it is meaningless to obey the provisions of western organization supervision, which makes the World Energy Charter almost useless.

The United Nations Framework Convention on Climate Change (UNFCCC) is not effective. The protocols, including UNFCCC, were simply regarded as proposals

---

[*] See *Introduction to New and Renewable Energy*, Wang Gehua, Chemical Industry Press, 2006.

at the Copenhagen Summit in December 2009. G8* has been discussing climate and energy problems almost every year for the past 10 years, but has not taken any action. The special forum held in London by G20 for limiting greenhouse gas emission in October 2009 showed few achievements, neither did the related conferences after that.

Concerns on Clean Development Mechanism

Clean Development Mechanism (CDM) is one of the flexible implementation mechanisms introduced in the Kyoto Protocol (KP). Its core contents are to allow developed countries to implement the emission reduction projects helpful for the sustainable development of developing countries so as to reduce greenhouse gas emission and carry out the emission limitation or reduction duties promised by them in KP.

Yet, two-thirds of the supposed "emission reduction" credit limit produced by CDM in developing countries' projects does not really reduce pollution. The cost of the emission reduction realized by CDM is usually surprisingly high: Emission reduction through international funds not connected with CDM are seemingly effective and can save billions of dollars.† Besides, it is no good to the global climate if one Chinese mine reduces methane emission under CDM because the polluter that buys compensation evades its own emission obligations.

Many people hoped that CDM would promote the development of renewable energy and improve energy efficiency. However, if CERs until 2012 have been obtained for all projects that are being built, non-water-renewable energy will attract 16% CDM capital, and the energy efficiency projects on the demand side only 1%. Only 16 solar projects—less than 0.5% of the projects that are being built—have applied to CDM for approval.

CDM provides opportunities for developed countries to conduct emission reduction and investment in developing countries. Under the mechanism, the developed countries

---

* G8 refers to the economic cooperation mechanism between the eight industrialized powers, America, Britain, France, Germany, Italy, Canada, Japan, and Russia; it was later developed into G20.

† CDM credit limit is a Certified Emission Reduction (CER), representing 1 ton of carbon dioxide that is not emitted into air. Industrialized countries' governments buy and use CERs to prove to the UN that they are carrying out their emission reduction duties stipulated in the KP. Companies also buy CERs to comply with national laws or EU's ETS (emission trading scheme). It is estimated that two-thirds of the emission reduction duties of the main developed countries that signed KP can be implemented through buying compensation, not letting their economy be decarbonized.

need to use their benefits for the sustainable development of the countries where the projects are. Some people criticize the CDM saying that many CDM projects neither settle the real problems that industrialized countries face in greenhouse gas emission reduction nor promote the sustainable development of the countries where the projects are. The CDM projects were meant to embody the developed countries' pursuit of clean coal and other low-cost projects. However, the developed countries think that the renewable energy projects of developing countries are huge in investment cost and low in return rate, so they are not interested in them.

CDM is also not helpful for promoting energy efficiency and vehicle efficiency, which are vital for developing countries' emission reduction and sustainable development. The World Bank (WB) estimates that there is large room for improvement in global energy efficiency. But, limited global projects of energy efficiency improvement fail to fully embody the meaning of energy saving and emission reduction. WB still believes that CDM has not attained its potential of promoting world sustainable development that it should have.[*] More and more evidence shows that CDM increases greenhouse gas emissions under the cover of promoting sustainable development just because of vested interests.[†]

Smart Energy: the Light of Our Future

As humans went through the process of civilization, energy forms kept improving and updating, revealing the continuous evolution of human intelligence. As we adopted more ecologically sustainable energy forms, a new concept—Smart Energy— began to enlighten and indicate the direction of improvement and transformation for future energy. Smart energy, as a clean, highly efficient, and sustainable energy form, will be able to relieve the energy supply pressure, promote and smooth the transformation to an ecological civilization, and promote the vigor of future civilizations.

---

[*] See *Future of Energy—Low Carbon Transformation Route Map*, Phil O'Keeffe, et al., Petroleum Industry Press, 2011.

[†] Adapted from McCully, P. Discredited strategy, *The Guardian*, May 20, 2008. https://www.theguardian.com/environment/2008/may/21/environment.carbontrading.

Chapter 16

Evolution of Civilization: Driven by the Changes of Energy Form

16.1 Evolution of Civilization Forms

The word "civilization" originates from modern Europe. It was originally used to describe human behavior. Those who are educated, polite, and enlightened were called civilized people. Later, it also came to be used as a measure of social development, describing the level of social progress, reflecting the state and development trend of social evolution, in contrast to barbarism.

Mankind has experienced the prehistoric hunting–gathering civilization, the ancient agricultural civilization, the modern industrial civilization, and the contemporary information civilization. It is hoped that the future civilization would be an ecological one marked by the wide adoption of smart energy (see Table 16.1).

Despite the fact that the hunting–gathering civilization period lasted the longest (about 3 million years), its social productivity level was the lowest. Humankind directly obtained resources from nature—gathering plant fruits and hunting animals. In this period, our ancestors also learned to use rough stoneware to enhance their hunting–gathering efficiency, invented fire-making technology to widen food sources and regional scope, and created bows and arrows to improve their defense ability and enrich their means of livelihood. In summary, humankind was extremely short of materials in the hunting–gathering civilization and almost completely relied on nature for food; it was then that the spiritual culture had appeared.

As means of livelihood was enriched, settling down gradually became the norm; the population gradually shifted from gathering and hunting to agriculture,

Table 16.1   Different Human Civilization Forms and Energy Use Forms

Development Stage	Time	Energy Used	Main Energy Forms
Hunting–gathering civilization	About 3 million years ago to 12,000 years ago	Use of fire	Firewood
Agricultural civilization	About 12,000 years ago to the 1500s	Use of domestic animals, wind wheels, and water wheels	Animal power, wind power, water power
Industrial civilization	1500–1945	Use of steam engine and internal combustion engine	Coal
Information civilization	1945–now	Use of power generators	Electricity, petroleum, coal, natural gas
Ecological civilization	Future	Wide adoption of smart energy	Smart energy

and obtaining means of livelihood mainly became manpower, animal power, wind power, and water power. Agriculture became the main driving force of social development. We entered the agricultural civilization relied on then, and the period lasted over 5000 years (some people believe it is about 10,000 years). In this period, accumulation of material and wealth was relatively slow because of the restraint of limited land resources and productivity level, but the scale and scope of population migration far exceeded the hunting–gathering civilization; division of labor was gradually brought in; culture, art, politics slowly formed; and cities, classes, and states appeared. The social form of alienation of the human nature appeared in the course of social scale expansion and material civilization progress, and the slave system came into existence. Later, mankind became ideologically enlightened; reason and spirit were encouraged, human nature was emancipated, and the slave system gradually collapsed, only to be replaced by the dominant feudal society. Apart from the ruling class, people basically could equally coexist and had independence and legal property. However, the social system was still unequal because of the ruling class. Religious thought still dominated human thinking at that time, and the world's three main religions (Buddhism, Islam, and Christianity) played a very important role in the ideology of the civilization of each region. In summary, material and wealth were relatively sufficient in this period, but the dependency on nature was also relatively serious; the development of culture and art was relatively fast, but their overdevelopment hindered economic development and social progress to some extent; humanity was liberated to some extent, but political inequality was still prominent.

In the later period of the agricultural civilization, the commerce and handicraft industry were developed, and the market gradually expanded. Industry and commerce gradually replaced agriculture as the mainstream livelihood, and the beginning of another civilization was marked by the use of steam engines. The increase in productivity and improvement of medical and health conditions made the population grow faster and migrate from the countryside to cities. Cities increased in size and citizens made cities the center of economy, finance, politics, culture, and education—it became the "locomotive" for the development of industrial civilization. At the same time, the rapid development of economy and society made us undergo an unprecedented change.

The industrial civilization increased the fortune of humans and led to the formulation of the concepts of humanity, liberation, freedom, and equality; these concepts found their way deep into the minds of people. However, this civilization also brought many problems, such as frequent wars, environmental pollution, and moral depravity; if it had been allowed to flourish unchecked, it is extremely likely that it would have led human civilization into an abyss. Therefore, subsequent generations of people faced two basic problems: One was to further develop the existing industrial civilization; the other was to avoid taking the old road of the industrial civilization period. The core of the first problem was how to find new economic growth points. The industrial civilization brought a rapid flow of talents, capital, technologies, and commodities to the forefront; information was the only major field left untouched as it was very complicated—how to create, store, use, recombine, and recreate information resources was an urgent problem that needed to be settled. The core of the second problem was how to avoid the overconsumption of resources and energy and how to halt the environmental damage brought about by economic growth.

We began to vigorously develop information technology, represented by computers and advanced new technologies such as communication, microelectronics, photoelectricity, sensors, and superconductors. Their wide adoption marked the fact that mankind had entered a new era—the information civilization era. Different regions, states, and nations vary in their economic and social development. Some have entered the information civilization era, some are entering it, and some are still in the era of traditional industrial and even agricultural civilization. So, the demarcating line between the information civilization and the industrial civilization is difficult to determine. Besides, different people call the information civilization differently, some call it the "postindustrial era," some call it "the third wave," some call it the "knowledge economy" or the "low carbon economy," and so on. Whatever its name, the features are basically the same, namely, a civilization dominated by science, technology, knowledge, culture, politics, society, and other kinds of social civilizations. Some scholars further subdivide it into the computer revolution period, the information revolution period, and the organic revolution period.* We can more deeply experience life compared with the industrial civilization; the information civilization is smarter, cleaner, lower in carbon emissions, and more efficient; thus, this civilization can better embody the integrated development of human civilization (both material and spiritual).

As mentioned earlier, the technology, politics, society, and culture of each civilization are unique, but the later form is the previous civilization form's extension and evolution. The agricultural civilization was more advanced and prosperous than the hunting–gathering civilization because they used advanced tools for manufacturing and living (such as stoneware, metalware, and chinaware). Of course, in the agricultural civilization period, the productivity level was still relatively low, social division of labor was simple (people were not all considered equal), and the superstructure hindered the social progress and improvement of productivity level in its later period. The industrial civilization remedied the aforementioned disadvantages; productivity level was greatly improved; material properties were greatly increased in amount; and democracy, freedom, peace, and other ideals were recognized by more and more people. But, the industrial civilization also has its disadvantages, such as resource shortages, environmental pollution, climate damage, disputes between nations and regions (even world wars), inequality between states, regions, people, and industries (colonies were a miniature model of the inequality between states). The contemporary information civilization and the mixed civilization of the transitional period have partially solved these problems, such as the appearance and use of mass production, which made life more convenient and reduced energy consumption; the application of networks and other media has changed the social structure greatly; national independence movements rose worldwide, and the colonies established during the industrial civilization disintegrated.

Even so, the contemporary information civilization has still not been able to solve all problems such as conflicts over resources, environment, energy, population, society, and politics, which, hopefully, will be solved by future civilizations. The settlement of these problems is the very reason why the future civilization will be more advanced than the information civilization. Although there is still controversy in the cognition of the future civilization, it is generally believed that we are heading toward an ecological civilization. The future ecological civilization will be superior to the contemporary information civilization in the following aspects. First, the tools of the ecological civilization will be more advanced. The production tool is the mark that shows whether a civilization is more advanced than another. There is nothing more advanced than the ecological civilization. It

---

* See *Future of Mankind,* Zhu Zhongwan, Liaoning Science and Technology Publishing House, 2010.

gathers all advantages of the information civilization and overcomes its limitations in technology and other aspects. For example, at present, network technology is widely used, but network nodes are limited. The ecological civilization will further expand the width and depth of the Internet. Network technology will profoundly change human activities, and other technologies (such as nanometer technology, sensing technology, photoelectric technology, acoustoelectric technology, and new material technology) will also undergo profound changes. Besides, the material basis of the ecological civilization is likely to change; compared to inorganic substances, the more intelligent organism will more and more become the material basis of the ecological civilization. Second, the system of the ecological civilization is more reasonable. Technical progress and global allocation of resources require the corresponding systems to change. For example, network information crossing borders greatly increases transparency, makes the public's requirements and voice be responded to in time, and results in the change of social organizational structure and political system. Another example is the effect of global resources on the regional system; the global allocation of resources will impact the regional (national) system; the overall system will adapt to global and regional allocation of resources, and economic and social system (policies, system, and mechanism) will be changed and optimized accordingly.

East and West: Civilizations with Different Characteristics

Civilization refers to the sum of the wealth mankind has created in a broad sense and the spiritual wealth in a narrow sense, such as literature, art, education, science and so on. It covers the relationship between people and people, people and society, and people and nature. A civilization form is the values and norms that people follow to be happier, better, and more harmonious. There is a significant difference in civilizations in different regions, and this is affected by history, culture, ethnic groups, religions, natural environment, regional distribution, and other factors. There are various theories about how to classify civilization. Civilization can be roughly divided into east and west based on geographical distribution. The eastern civilization, taking the self-restraint and self-regulation characteristic of human nature as its starting point, emphasized the instructive role of morality; the western civilization, taking the liberation and freedom loving characteristic of human nature as its starting point, emphasized the compulsory role of laws.

The eastern civilization refers to the civilization of the eastern part of the world, which consists of countries in Asia and

North Africa. It is represented by Chinese, Indian, and Arabic Islamic civilizations. The Chinese agricultural civilization, characterized by self-sufficiency, was the most advanced in terms of scientific, cultural, and economic strength and had the highest income level in the world at that time. The Indian civilization with its unique values and ideological system occupies an important status in world civilization. The Arabic Islamic civilization stuck to pure concepts and pursued lofty ideals. In short, the eastern civilization paid attention to apperception, training, restraining desires and self-denial, moderation, excessively complying with nature, and advocating gradual changes to a higher ideal.

The western civilization originated from ancient Greece and ancient Rome with infertile land surrounded by water in three directions. The ancient western civilization gradually declined because of the gap between the rich and the poor, internal conflicts, natural disasters, etc. Centuries later, the Renaissance ushered in western modern history. The two Industrial Revolutions originated in the west and brought in the world industrial civilization; there was a sudden expansion of wealth and market demand. Because western countries are generally small and could not meet the development requirement of the industrial civilization, the western civilization gradually opened up and expanded to the outside world. The western civilization paid more attention to logical reasoning, experiments, invention, opening up, and freedom; they promoted technical development, rapid wealth accumulation, and cultural and artistic recovery and improved people's living standards. Since the Industrial Revolution, the western civilization has influenced and even dominated world development and international order. But it has also brought a series of problems: expansion of private desire with no thought to expense, extravagance and degeneration, advocating the "jungle law" and abusing military force.

The eastern civilization and western civilization have their own characteristics, disadvantages, and advantages, and they both have made great contributions to the progress of mankind. As information, traffic, and other technologies develop, high mountains and vast seas cannot stop the communication and cooperation between the east and the west; the eastern civilization and the western civilization must gradually integrate. However, the biggest obstacle at present is the inertia of our thinking and our resistance to change, which sets up a fence between our hearts. The future civilization, which is on its way, is no longer privately owned by one region or nation and cannot be created by any one region or nation. It is not necessary for us to dispute who is right or wrong, or strong or weak—the eastern civilization or the western civilization. We should expand our horizons to include the whole world, reach a consensus, absorb the essence of all kinds of civilizations, and eliminate difference and conflicts together. We should create a new civilization that serves the whole of mankind, instruct mankind to eliminate poverty, war, chaos, and pollution and go together toward a happier, more harmonious, greener, and better future.

16.2 Transformation of Energy Forms

The energy forms we use can be divided into initial smart energy, lesser smart energy, medium smart energy, and greater smart energy; and their corresponding energy forms are respectively firewood fuel, "domesticated" energy, fossil fuel, and hybrid energy.

In the period of the gathering and hunting civilization, we obtained food from nature, gathering plants, and hunting animals to get energy for survival. Later, we found and used fire; making firewood, straws, and other natural biomass fuel became our main energy.

In the period of agricultural civilization, in addition to the use of fire, we began to attempt to "domesticate" nature. Animal power, wind power, and water power gradually became important modes of obtaining energy and helped us finish simple agricultural and handicraft activities, such as grinding flour, lifting water, spinning yarn, and weaving cloth. We now believe that there are obvious defects in directly obtaining water power, wind power, and other energy forms because of the following reasons: (1) energy power is greatly limited by natural conditions and climate, and windmills and water wheels cannot rotate without wind and water; (2) the approaches of obtaining energy are very limited, mainly through the earth's surface and above the surface. The above-the-surface energy is solar energy. The earth's surface energy is the energy formed by solar energy through photosynthesis, such as firewood. At that time, we did not have the technology to tap underground energy. Because the production technology level was backward, our demand for energy was not so urgent; our driving force for improving energy-using technologies was insufficient. Low energy technology and low energy demand conformed to each other, interacted as both causes and effects, and resulted in the civilization going forward slowly.

Benefiting from technical progress, our activity scope gradually spread from the surface to underground, where we discovered rich coal resources. Coal could release relatively high heat after combustion, suited mass production, and met demand quickly; therefore it became the main source of energy. The invention of the steam engine in the 18th century marked the point when coal replaced firewood as the main source of energy; coal supported the industrial civilization. As the internal combustion engine with higher thermal efficiency, smaller volume, and bigger and cleaner power appeared in the 19th century, we entered the times of petroleum as the main fuel. Petroleum and natural gas, which have higher combustion value, less pollution, and can better fit the demand of the industrial society, were massively extracted and used.

The power produced by the steam engine and the internal combustion engine was difficult to transmit freely over long distances; we could only conduct small-scale products over small distances. The discovery of the law of electromagnetic induction that provided the basis for the subsequent invention and wide application of a series of technical equipment, such as generators, electric motors, transformers, power stations, low-voltage grids, and extra-high-voltage grids, started the second Industrial Revolution and ushered in new electromagnetic power. The extensive demand of urban and rural life was met by coal, petroleum, water power, and wind power that could be converted into mechanical energy, which was then converted into electric energy and transmitted to cities and villages through the grid.

We changed from relying on manpower, animal power, wind power, and water power to relying on machines by dint of the steam engine and the internal combustion engine using coal and oil as fuel. Human's power over the world expanded outwards. We converted all kinds of mechanical energy into electric power by virtue of the discovery and application of electricity and transmitted electric power to far off places, making our production higher and our life easier. With the discovery and application of fossil energy—the energy basis of the two industrial revolutions—the huge energy demand of industrial civilization was satisfied. But fossil energy is limited, nonrenewable, and will ultimately get depleted. Meanwhile, it also causes a series of serious problems such as environmental damage, climate change, and safety accidents.

The energy situation for future development is grim, and clean, highly efficient new energy sources are indispensable. Since the 1960s, the voice of the "energy revolution" is increasingly getting louder. New energy and renewable energy has become the new goal of energy form replacement and development.

Signs of How Energy Forms Will Be Replaced

Energy forms have to be improved if human subsistence is to improve and civilization is to progress. The replacement of energy forms pushes ahead human development and the evolution of civilization forms. The replacement cannot occur suddenly; it requires the improvement of energy forms. Both the improvement and replacement of energy forms follow the successful use of human ingenuity and wisdom. Looking back to the long development course of civilization starting with fire, we can find several clues or signs of the improvement and replacement of energy forms.

From passive to active. In the earliest times, we obtained energy in the passive way "depend on God for food," namely, gathering plants and hunting animals to obtain the energy they contained. Later, we changed from passively using energy to actively using energy through using fire, including using natural fire and drilling wood to making fire. With ceaseless technical progress, our energy vision was gradually transferred to coal and petroleum. Nuclear energy, solar energy, wind energy, water energy, and other new energy forms are all examples of our actively using energy.

From discovery to invention. Tens of thousands of years passed from when we used fire left by volcanic eruption and lightning to when we began drilling wood to make fire. But there was only a 11 year time period from when Hans Oersted* discovered the magnetic effect of electric current to when Faraday produced electric energy with a magnetic rod passing in and out of a metal coil. Accidental discovery is an event with low probability; one needs to wait for a long time, and even then only relatively little information can be found. Experimentation and invention are complicated systematic processes; but they have a clear purpose and research direction; so they could greatly shorten the times between inventions. ·

From shallow to deep. In the periods of the hunting–gathering civilization and agricultural civilization, we gathered plants and hunted animals that we could see and obtained energy from them. In the period of industrial civilization, underground coal, petroleum, and natural gas became important power sources for the advancement of civilization, but these forms of energy still could be seen with eyes. With technical progress, energy forms have changed from visible to invisible; electric energy was universally used; and nuclear energy, solar energy, flammable ice, and geothermal energy entered the gambit.

---

* Hans Oersted (1777–1851), a Danish physician, chemist, man of letters, first discovered the magnetic effect of electric current.

In short, use of energy shows our progress from lesser to greater intelligence. People used animal power and transported goods by horse-driven carts in the period of the agricultural civilization. The steam engine, internal combustion engine, electric generators, and electric motors enriched and strengthened energy. Later, hybrid energy was used. Later again, energy forms were further upgraded. Fuel cells, hybrid power, hydrogen energy power, and solar energy were rapidly developed. With the evolution of civilization and improvement of productive technology level, mainstream energy forms will contain more properties of human intelligence.

16.3 Relationship between Civilization and Energy

So far, we have moved past the hunting–gathering civilization, agricultural civilization, industrial civilization, and information civilization. The evolution of human civilization forms is closely related with the replacement of energy forms, and they promote and complement each other. The replacement of energy forms promotes civilization evolution, and the civilization evolution pushes the replacement of energy forms in turn (see Figure 16.1).

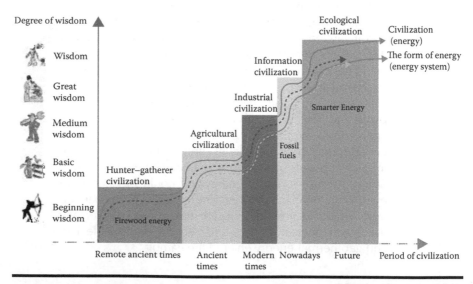

Figure 16.1 The curve of the replacement of energy forms and civilization evolution.

The hunting–gathering civilization and initial intelligent civilization. In the period of the hunting–gathering civilization, where gathering plants and hunting animals was the basic way of life, we were not greatly different from animals. But there was a difference between us and other animals in nature: mankind could use primary energy. We discovered fire in practice and used it for driving away cold, protecting ourselves, capturing prey, domesticating animals, boiling and cooking food, and storing seeds, thereby making us superior than animals; even then we dominated the earth. We spend the long period of the hunting–gathering civilization dependent on fire.

Agricultural civilization and lesser smart energy. After we had learned to cultivate plants and domesticate animals, agricultural settlements became possible, and we changed from plant gatherers to plant producers—we had entered the agricultural civilization. Textual researches show that the agricultural civilization zones were mainly between 20° N and 40° N latitides. It can be said that these regions are the birthplace of early human civilization. In this period, most people were peasants who lived by cultivating plants and domesticating animals; the rest were engaged in related works. Then division of labor appeared, and classes and states were formed. We burned plants to produce thermal energy and used domesticated animals, wind power, and water power to produce mechanical energy. The period lasted thousands of years and provided limited power, and productivity was low; it was difficult to transport people and goods to far off areas, surplus means of livelihood was less, people's activity space was limited, their thoughts were conservative, division of labor was simple, and the majority of people were engaged in agricultural labor.

Industrial civilization and medium smart energy. With more surplus means of livelihood, subdivision of labor, and frequent commodity exchange, goods needed to be transported to more distant areas. The scope for trade gradually expanded; we desperately needed more energy-intensive power. Our advancing steps were quickened by the discovery and use of coal. Compared with plants, coal was high in combustion value and could provide more power. The steam engine was put into industrial production in the 18th century, widely used in all kinds of production fields in the early 19th century, became the "locomotive" of the Industrial Revolution, and made the massive use of coal possible.

Information civilization and greater smart energy. The appearance of the internal combustion engine promoted the development of the transportation industry. Petroleum was widely used. It was both a high-quality raw material of power fuel and an important chemical raw material. Besides, people used the stable and easily transmitted electricity produced by generators to drive large-sized electric motors, which further promoted the development of the transportation industry. Petroleum and electricity became the strong driving force of economic and social development. And the development of the industrial civilization was greatly accelerated by the energy that was represented by primary energy, such as coal, petroleum, natural gas, and secondary energy, such as electricity that is used in the functioning of motors, internal combustion engines, cars, planes, rotating furnaces for making steel, telephones, and radio communication devices. With energy-intensive power, people could transport goods to each corner of the world, which greatly promotes trade development. The massive mining and use of those resources became real, providing raw materials in an endless stream for industrial production. Massive agricultural mechanization brought prosperity to agricultural production. With more convenient traffic, the rural population rapidly migrated to cities, further promoting evolution of civilization.

The industrial civilization, which lasted for over 100 years, dramatically enriched humans. But problems arose one after another: environment pollution; climate change resulting from industrial activities and possibly challenging our subsistence; and energy and resource shortages. Although the industrial civilization created tremendous wealth, it consumed treasured resources and energy, most of which were nonrenewable. If we persist along this road, the industrial civilization will be unsustainable, and mankind will go into a cul-de-sac.

In the period of information civilization, both energy and resources are saved. But the main energy is still fossil energy, which will be depleted in the end; so the evolution of civilization is certain to be restrained by resources and energy. Therefore, we need to develop smart energy soon. It will promote civilization evolution step by step through the continuous improvement and replacement of energy technologies and usher in the intelligent epoch of energy use.

"Double P" Principle

The "double P" principle, namely, push and pull principle, was first proposed by D.J. Bagne. It is an important theory explaining the population migration phenomenon. The theory argues that population flow is the result of the push from outflow areas with disadvantageous conditions and the pull from inflow areas with advantageous conditions. Bagne believes that the purpose of population flow is to change the original living conditions—the disadvantageous living conditions of outflow areas become the push, and the factors that are helpful for improving living conditions are the pull.

The "double P" principle can similarly explain the push–pull interactive relationship between energy and civilization. On one hand, energy supports and pushes human civilization evolution, and different energy forms support different civilization forms. In the period of the hunting–gathering civilization, we only used the energy from the natural world or other mechanical energy to meet the demand of civilization. In the period of the agricultural civilization, we used labor power, animal power, wind power, and water power and expanded the activity scope of civilization. In the period of the industrial civilization, the wide use of coal, petroleum, natural gas, and electricity pushed the civilization to a new height, and productive and technical level was further improved.

On the other hand, the evolution of civilization puts forward new requirements for energy forms. In the period of the hunting–gathering civilization, we performed activities for survival, contending with nature, especially with animals, obtaining food and staying warm. Hence, we only needed to use fire to drive away animals and cook food. There was no need for us to consume too much energy to go to faraway places. In the period of the agricultural civilization, surplus products gradually appeared, the forms of energy were diversified, social division of labor was created, and agricultural produce was gradually increased in amount and required, necessitating transport to other places. People began to explore better energy forms to meet activity demand at the end of the period. In the period of the industrial civilization, the activities of production and trade were rapidly expanded; the demand of means of transportation, production, and communication, etc., was enhanced; energy helped people and goods to be transported to other places; and the energy forms, such as coal, petroleum, natural gas, and electricity, satisfied the energy needs of human activities back then. Of course, these energy forms based on fossil energy had many problems, such as limited supply and damaging the ecological environment. It can be seen that the new civilization form will put forward higher requirements in energy safety, cleanness, low carbon, environment protection, high efficiency, convenience, and so on.

Chapter 17

Smart Energy: The Wonderful Production of Time

17.1 Basic Connotation of Smart Energy

In 2009, IBM experts deemed that the contemporary world, in which economy, technology, society, individuals, and units were all mutually interconnected, had really become an interconnected network world and that the earth had become more flat and smaller. The world would become smarter if intelligence was injected into our work system and methods. They believed that interconnection of science and technology would change the way the whole human world operated and would be involved in the life and work of billions of people. Hence, they innovatively proposed to "build a smarter planet" and presented many concepts, such as smart airports, smart banks, smart railways, smart cities, smart electricity, smart grid, and smart energy, which would form the Internet of things through universal connection. The Internet of things is integrated through supercomputers and cloud computation to make people manage production and live in a better and more dynamic way so as to attain a global intelligent state and finally achieve "Internet + Internet of things = Smarter planet." IBM as an enterprise put forward a series of intelligent concepts, but it did not break through the limitations of commercial goals and the scientific point of view.

In the same year, Cleantech China's Chief Information Officer Han Xiaoping wrote *When Energy Is Full Of Intelligence, Intelligent Energy and the Progress of Human Civilization* and other articles, which attracted domestic attention on smart energy. The concept took root in China from then on. At present, people are still not accustomed to linking the term "smart" with energy; instead, they prefer to use the word "intelligent energy," which tends to refer more to the technical aspects of energy. Although both the terms "smart energy" and "intelligent energy" are often used in journal articles, there are still no authorized definitions of them.

To adapt to the new trend and requirement of civilization evolution and protect itself, mankind must resolve the power problem of civilization evolution and realize energy safety, stability, cleanness, and perpetual use. The essence of smart energy is to develop mankind's intelligence and ability, integrate human intelligence in the course of energy development, utilization, production, and consumption through continuous innovation and system change, and establish and improve energy technologies and systems conforming to ecological civilization and sustainable development requirements so as to set up a new energy form. In short, smart energy is to have self-organization, self-examination, self-balance, self-optimization, and other human brain functions to meet the demands of system, safety, cleanness, and economy (see Figure 17.1). It is easy to affirm that smart energy will be a new milestone in the history of how humans use energy.

The carrier of smart energy is energy. The object and carrier we study is always energy, whether in development of utilization technology or in the production consumption system. Our purpose of unremitting exploration is to look for safer, more sufficient, and cleaner energy to make human life happier, commodities cheaper and finer, and ecological environment more beautiful and suitable for living.

The guarantee of smart energy is systematic organization. For the smart energy that will bring new energy patterns, an advanced system is required; this is necessary for providing relative guarantees that can encourage scientific innovation, optimize industrial organizing, advocate energy saving, promote international cooperation to ensure stable operation, and accelerate development of a smart energy system.

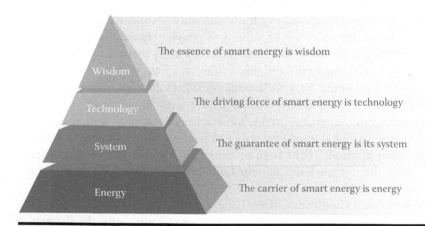

Figure 17.1 Factors of smart energy.

The power behind smart energy is science and technology. The scientific innovation of the steam engine and internal combustion engine promoted the development of the industrial civilization. Similarly, the development of smart energy also needs science and technology to promote it. Nuclear energy, solar energy, biomass energy, the ubiquitous network, and other energy-developing and -using technologies that we are using, exploring, or have not invented yet will provide huge clout for the development of smart energy.

The essence of smart energy is wisdom. Wisdom is the ability to know, distinguish, judge, dispose, invent, and create things. The wisdom of smart energy is not only integrated into technical development, use, and innovation of energy but also embodied in energy production and consumption system change.

Smart energy is not simply equal to smart energy technology; it should also include a smart energy system. Technology is the fundamental power of smart energy development, and its system is the fundamental guarantee of smart energy development. Both are indispensable. Smart energy not only refers to energy developing and using technology; it also includes the energy production and consumption system. Technologically, smart energy not only refers to the innovating technology of traditional energy but also includes the developing and utilization of new energy. Systematically, smart energy not only refers to the energy production and consumption system but also includes all systems related to energy. In time, smart energy not only refers to the innovation of present energy technology and the improvement of the present energy systems; it also includes the development and utilization of new energy forms and technologies and systems related to human man life.

There are connections and important differences between the concepts of smart energy, new energy, renewable energy, and clean energy.

Smart energy combined with energy efficiency technology and intelligent technology will emphasize specific technology and its physical and material attributes.

New energy is a kind of energy distinct from conventional energy. Its prominent features are its technological advancement, incomplete commercialized development, and scale use, and some examples of this include wind energy, solar energy, marine energy, geothermal energy, biomass energy, hydrogen energy, and nuclear fusion energy. In term of energy forms, new energy belongs to smart energy, but the extent of smart energy is larger than that of new energy. The clean and high-efficiency technology of traditional energy also belongs to the category of smart energy.

Renewable energy, relative to nonrenewable energy, emphasizes the renewability of energy under certain time and space. It, no doubt, is part of smart energy because it realizes the renewability of energy and embodies human intelligence. But renewable energy is not equal to smart energy because the scope of smart energy is far wider than renewable energy. The technological innovation of nonrenewable energy also belongs to the category of smart energy.

Clean energy is also not equal to smart energy. Though cleanness is an important attribute of smart energy, all clean energy is not included in the category of smart energy. Clean energy can become smart energy only when it satisfies the condition of high efficiency, safety, etc. Therefore, clean energy and smart energy intersect but do not completely coincide.

New Land of Peach Blossoms

A Wuling man left the Land of Peach Blossoms and marked the way. When he arrived at the place where the governor resided, he visited the governor and informed him that he had marked the way to the Land of Peach Blossoms. The governor immediately sent people to go with him. But when they looked for the marks, they lost their way and could not find their way back. When he was about to die, the Wuling man drew the picture of the Land of Peach Blossoms and gave it to his sons. His sons gave birth to grandsons, who gave birth to children, and so the picture was passed on from generation to generation. In this way, over 2000 years passed.

One day, a descendant of the Wuling man set out on a journey looking for his ancestors. He saw a stone tablet much like the one described on the picture drawn and passed on to him by the Wuling man. He climbed it, saw a flowing stream and peach blossom petals floating serenely in it; it delighted him very much. There was a mountain at the end of the stream. The descendant entered the mouth of the mountain.

There in the mountain was an old man hunkering on a big stone. The descendant told the old man the stories about his ancestors and the Land of Peach Blossoms. The old man laughed and said the place was originally the Land of Peach Blossoms, but now it was called the Prefecture of Intelligence. The descendant was confused and asked, "What is intelligence?" The old man replied, "Intelligence is ability, schemes, wide vision, unique thought, food, clothing, shelter and transportation, ethics, morality and every phenomenon under the sun, and the fountain that moistens states and people."

Following the old man, the descendant entered the prefecture, which was crowded with people and vehicles, very prosperous, and with no black and foul atmosphere. People came and went, cheerfully. No noisy sound could be heard. A huge apparatus could be seen, run in space and underground, travelling 10,000 miles in a second. It was intelligent traffic. There were many very healthy centenarians. Several toddlers with only tufts of hair on their head were asked to recite the poems of Tang and Song dynasties, and they did so fluently; they were thorough in astronomy, geography, and all fields of knowledge. The descendant exclaimed at the miracle of intelligence. At dusk, there was no chicken, duck, fish or other meat at the banquet; green leaves and fruit were eaten. The descendant was full and felt energetic. People traveled to and fro between celestial bodies by virtue of the power of solar wind. The descendant knew that there were also states and families outside earth. At night, lights illuminated the darkness like sunlight. One did not know the daytime had passed if he did not have a watch. The bed was floating in the air; sleeping in it was like sleeping in a cradle. It was winter, but a warm wind was blowing and the cold was not felt. It was difficult to distinguish summer and winter as there was no calendar. The descendant could not understand how this miracle was possible. A learned man cleared up the matter in a single sentence. Smart energy is mysterious and magical, endless, uses fewer resources and produces immense power, can be transported to very far places in less than a second, and produces little pollution and residue. With plenty of energy, people do not need to worry about haze or fight for coal and petroleum.

The descendant asked whether smart energy could be passed on to outsiders. The answer was yes. He could learn the technology of smart energy if he was willing. He stayed there for months and left after he had learnt the technology thoroughly.

Having left the Land of Peach Blossoms, the descendant received apprentices to teach them the technology, and thus left the fragrance of peach blossoms in the world.

17.2 The Importance of Smart Energy

Smart energy that is a child of the present time is the light of human civilization's future. To develop and use smart energy will not only ease the energy crisis and pressure and satisfy the need of the present and future development of human civilization but also promote civilization evolution and speed up civilization transformation.

Relieving realistic pressure. Smart energy helps us to relieve our pressing need and reserve wider space, thereby providing more time for the development of human civilization evolution.

1. *Relieving environmental damage.* Smart energy is clean energy. It produces little waste material in production, transmission, and consumption and can help to effectively control noise, radiation, and thermal pollution; it can fundamentally eliminate the adverse changes we had brought on to geology, hydrology, and the atmosphere and can restore the natural environment through its own ability to realize balance.

2. *Relieving resource shortages.* We have consumed large amounts of fossil energy, which has resulted in fuel shortage, the surge of production costs, and increase of consumers' cost of living. Shortage impacts economic operation and leads to global economic decline, which results in economic and social problems. Smart energy promotes the massive and commercial use of new energy and guarantees sustainable, safe, and stable energy supply through improvement of all kinds of energy systems, innovation of energy technology, reduction of energy consumption, and reduction of the production and consumption cost of new energy.

3. *Relieving energy disputes.* The energy shortage and environmental damage caused by energy consumption imposes pressure on each country and region, leads to ceaseless energy disputes, and even wars. It is easy to see after analysis of human history of over 100 years in the past that contending for interests, especially energy and resources, is one of the key factors resulting in war eruption. Smart energy helps to bring to the fore the fact that all countries seek common points while reserving differences and reducing conflicts through international organizations, politicians, entrepreneurs, all kinds of nongovernment organizations and folk groups to establish all kinds of energy organizations.

Acceleration of civilization transformation. At present, we are transforming from the information civilization to the ecological civilization. "The way stretches endless ahead, I shall search heaven and earth," a line from a Poem by Qu Yuan (c. 340–278 BC), who was a Chinese poet and minister and lived during the Warring States period of ancient China. The purpose of smart energy is to choose a clear goal, right direction and suitable path, accelerate the improvement and replacement of energy forms and shorten the course of transformation toward ecological civilization form.

1. *Acceleration of productivity level improvement.* Smart energy means the improvement of energy technology, including improving the efficiency of traditional energy and production, using clean energy, or using alternative new energy technology. The technology itself is the most active enabler of increasing productivity. The technical innovation of smart energy will greatly enhance the productivity level and provide a solid material basis for civilization transformation.

2. *Acceleration of productive relationship improvement.* Smart energy requires smart systems, reform and innovation of existing energy systems, including using a suitable energy policy to reform energy systems, and mechanisms to

form a set of effective energy rules and international cooperation system so as to create the external environment in which the whole world would save energy and use clean, high-efficiency, low-carbon, and environmental friendly energy. This kind of energy relationship formed in the course of production and consumption is a part of a productive relationship, helps to perfect the productive relationship, improve superstructure, and lay firmer foundations for social development.

3. *Realization of the benign interaction between productive force and productive relationship.* On one hand, smart energy that represents the innovation of energy technology requires adjustments to the energy system accordingly. On the other hand, smart energy also requires that the energy system be initiatively changed to adapt to the need of energy technology innovation. The benign interaction between energy technology and system also promotes the interaction of productive force and productive relationship and accelerates civilization evolution and transformation.

Adaption to future development. Smart energy not only can play an important role in the relief of real pressure and acceleration of civilization transformation, but can also adapt to and meet the requirements of future ecological civilization development.

The ecological civilization puts forward the new requirements of energy technology and energy system. It can be predicted that in the period of the ecological civilization, society will continue to develop, population will continue to increase, the information and intelligence level of production, service, and consumption of material products will be enhanced in an all-round way. To minimize the consumption of natural resources and energy, reduce and even eliminate environmental pollution, we must make more efforts to develop alternative technologies of new energy forms while continuing to research and develop innovative technologies of traditional energy forms. The production pattern of high consumption, high pollution, and low output is directly related to an uneven energy management system, wrong energy policy, nonscientific energy pricing mechanism, mismatch of energy encouraging, and restraining mechanism and other energy systems. Therefore, energy system change is the objective requirement of ecological civilization development.

Smart energy can adapt to and satisfy the new requirements of ecological civilization. It means that the innovative technology of traditional energy and the alternative technology of new energy are developed and applied. Their progress will reduce consumption, decrease (and even eliminate) pollution, and make energy supply safe, clean, and economical. At the same time, smart energy means system innovation and change, which helps to integrate resources and energy, improve their input–output ratio, and reduce the negative effects on environment and ecology. It can be predicted that safe, clean, economical, and systematic smart energy can adapt to and fully meet the demand of future ecological civilization.

Expectation of Human Body Energy Technology

We know that the invention and use of cooking sped up human evolution, enriched the human brain capacity, decreased the sharpness and length of teeth, related bony structure, and muscle tissues, and shortened intestines. From this, it can be inferred that the exploring channels and realizing modes of future energy technology will not be limited to the development of external energy technology. Nanometer, genetic, macromolecular, medical, biogenetic technologies, etc., are used to promote evolution in mankind's use of energy to change our clothing, food, shelter, and transportation, starting with the human body structure, organs, cell tissues, genes, and so on.

Clothing. At present, all kinds of clothing we wear are mainly used to keep us warm. In the future, we perhaps can develop "heat insulation cream" developed from the moisturizing factors in moisturizing skin care products that maintain the water content in the skin. We only need to smear a layer of heat-insulating material on the body to keep warm; little heat is radiated outward in winter. Humans will be unaffected by hot sunshine and air in summer, and so we can live a comfortable life—being warm in winter and cool in summer. Later in the future, we probably can improve the epidermis at the cellular level and make it have similar suitable heat-insulating functions, bringing on a revolution in the wearing of clothing.

Food. The heat of the human body comes from food. Generally speaking, a man of 60 kg standard weight needs to consume 1500–1600 cal heat a day in rest state and 1800–2000 cal heat a day if he has a medium level of activity. To meet such heat demand, a chimpanzee needs to spend 6 h chewing food a day, but mankind, who has learned to cook food, only need spend 1 h. But, mankind also needs to spend some time in cooking, and three meals a day are still troublesome. At present, there are innovative high-energy carbohydrate, fat, and protein food, such as milk tablets, compressed biscuits, and individual self-heating foods. American First Strike Ration (FSR) was developed in 2002–2004 and put into battlefield service in 2007. Its calories are considerable. One ration can provide 2900 cal of energy. But, only these are far from enough. It is believed that we will surpass cooking and will identify a new way of absorbing energy. At that time, eating will be more efficient and simpler. Whether delicious food will be needed is another matter.

Shelter. The core function of shelter is just keeping out of the wind and the rain. If we had a hard and portable shell like a snail, our dependency on houses would be likely to reduce greatly. Of course, we do not hope that scientific progress brings us back to the times of painful humble abodes. If there could be a kind of beautiful or invisible electromagnetic field protective shell that neither swords nor spears could enter and that was watertight and kept us warm, it probably could meet our demand to a large extent.

Transportation. Urban traffic jams are a real headache for people. But the narrow paths and winding mountain roads in remote mountain villages are much more inconvenient. Although mankind has made great achievements in domesticating power, whether trains, cars, planes, ships, bicycles or skidding of wheel, but all have limitations and risk and are not perfect. In the future, will the speed of the leopard be transplanted to human legs? Will angel's wings be attached to human shoulders? We have the responsibility to collectively look for a transporting means that is faster, safer, and can even go up into the sky and go underground.

Innovation is the power of human civilization evolution. We do not need to confine our thought and action in the present relatively stable and comfortable life. We must boldly think and imagine our unknown world and hope. These things that seem incredible at present may become true tomorrow because of the evolution of intelligence.

Chapter 18

Totally Natural: Integration of Technology and System

18.1 Technical Basis of Smart Energy

Energy technology is the essence of smart energy—the direct embodiment of intelligence. All kinds of technologies carrying intelligence are cells full of vitality of smart energy, forming the energetic living body of smart energy. The technologies of smart energy can be divided into two kinds: innovative technologies and alternative technologies. Innovative technologies mainly refer to clean technology, high-efficiency technology, and safe technology that target the development and use of traditional energy forms. Alternative technologies mainly refer to exploring, discovering, developing, and using technologies that target new energy forms.

There are two standards to distinguish innovative technologies from alternative technologies: Form and trend. Innovative technologies are existing traditional energy in form and the improvement and progress of making it cleaner, more efficient, and safer in trend. Alternative technologies are known and even unknown new energy and revolutionary future energy that can replace the existing main energy source and even fully meet human energy demand.

Innovative technologies and alternative technologies are like the two legs of smart energy; they coordinate, go side by side, and complement each other; neither can be neglected. They are periodic and transient, lay the technical foundation for alternative technologies, meet human demand at present and before the massive replacement of energy form; it emphasizes maintaining achievements. Alternative technologies are those that can be used long term and as the revolutionary source of main energy; these are found based on innovative technologies to replace the existing main energy source and support human civilization evolution for a long time; it emphasizes development. They are adapted to the development of society and civilization and cannot meet the demand of new society and civilization after they last a certain time, although they will develop to a certain extent and become innovative technologies. Therefore, we need to ceaselessly look for new alternative technologies.

How do we define smart energy technologies? Combining the energy forms replacement path and course mentioned earlier and the demand of present society development and future civilization, it is easy to arrive at the key characteristics of smart energy (see Figure 18.1).

System. Smart energy technologies are not a single technology but the combination of the present Internet, cloud computation, communication, and control with the new technologies of the future, which could provide comprehensive advantages in energy production, transmission, and utilization. The functions of the smart energy technologies are no longer just simple energy production, transmission, transaction, and consumption but the comprehensive system built by combining environment, society, humanity, politics, and other indexes.

Safety. Smart energy technologies must conform to the requirement of safety and ensure safe, stable, and sustainable energy for society while eliminating threats, such as fire, flood, lighting, and traffic accidents, which are brought about when the

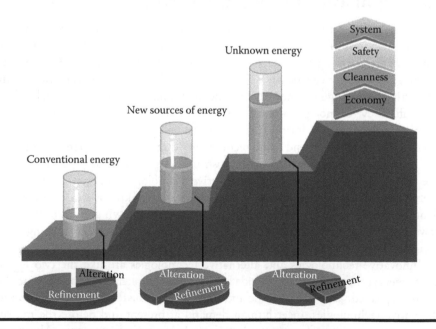

Figure 18.1 The development trend and key characteristics of smart energy technologies.

immense amount of energy contained in energy sources gets out of control, thus, thoroughly domesticating the "wildness" of energy.

Cleanness. The ecological civilization necessarily puts forward a very harsh requirement for the cleanness of energy. The effect of smart energy on the natural environment will be infinitely near zero, which is one of the final goals toward which we make unrelenting efforts. The clean attribute of future energy must be put at the forefront. There should be little or no harmful byproducts the course of production and use, and no effect on the ecological balance in the natural world. Smart energy not only strengthens the control of visible tangible pollutants but also eliminates the harm of radiation, electromagnetic waves, and other intangible pollutants.

Economy. With continuous improvement of the human intelligent attributes contained in energy technologies, energy efficiency will also be improved. Smart energy technologies will explore and develop higher efficiency energy, make it have higher and higher energy density, and lead to the obtainment of maximum power output with minimum cost. In short, it is highly efficient and low cost, provides high output, and is within the bearing capacity of the ecological environment and socioeconomic sustainable development.

Infinite Longing for Smart Energy Technology

The vast universe is full of the limitless unknown and holds infinite information. With the improvement of human science and technology and expansion of universe cognition in depth and width, we must discover and invent more new energy forms, some of which will perhaps be unexpected and some of which will even be incredible, far beyond our cognition scope but sure to bring wonderful magnificent future application prospects, just like science fiction. Combining frontier scientific research progress, we boldly envisage several future intelligent technologies, for example, "cold energy" technology. Until now, there has been no clear and authoritative definition of cold energy. The cold in "cold energy" signifies a relatively low temperature, namely, higher than absolute zero (-273.15°C) and lower than ordinary temperature (20°C), and its material carrier is a cold source, such as, ice, snow, and cold air. Cold technology is the technology of developing, collecting, storing, and applying cold sources to meet human demand. The potential advantages of cold energy are rich reserves in the natural world, complementary use with thermal energy, huge demand in daily life and industrial production, and so on. At present, our exploration and application of cold energy technology has just began; we cannot predict the extent of its future development.

Solar wind technology. Solar wind is a kind of supersonic-plasma-charged particle current that continuously exists. It comes from the sun and moves at the speed of 200–800 km/s. It is different from the air on the earth and does not consist of gas molecules but basic particles—protons and electrons—that are simpler and smaller than atoms. However, its flow is very similar to air flow, so it is called solar wind. The solar wind energy comes from solar flares or another climatic phenomenon called "solar storm," and its main characteristic is strong radiation.

Researchers from Washington State University in the U.S. plan to collect solar wind energy using a huge solar sail and satellite. It is estimated that 10^{27} W of power can be produced. If the electricity produced can be transmitted to earth, it can meet the electric demand of all mankind. However, it is meaningful only when it is transmitted to earth. Some of the power produced by the satellite is transmitted to copper wires to produce an electron collecting magnetic field, and the rest is used to provide energy for an ultraviolet laser beam to help to meet the electric demand of the whole earth under any environmental condition. However, the earth is too far from the satellite; the distance between them is millions of kilometers. Even the strongest laser beam will diffuse and lose most of its energy. Therefore, although most of the technologies used for developing this kind of satellite exist, an the development of a laser beam with a higher focus level is still a big challenge.

No doubt, there exists an immense amount of energy in solar wind, but the limits in practical operation will be a big problem; it can probably be of use to us in the future as human intelligence continuously increases.

Antimatter-using technology. Einstein believed all moving objects had energy. When its energy sum is a positive value, the matter is the matter we see in daily life. When its energy sum is a negative value, the nature and internal composition of the matter are contrary to those we see in daily life. This matter is called antimatter. The atom of matter consists of positively charged protons and negatively charged electrons, and the atom of antimatter consists of negatively charged protons and positively charged electrons. So, the direction of motion of antimatter after it is stressed is the complete reverse of that of matter. Matter and antimatter coexist with difficulty. They will attract each other once they meet and annihilate once they collide, releasing an immense amount of energy, which is called the annihilation reaction. In the annihilation reaction, matter and antimatter are both converted into huge amounts of energy, about 1000 times higher than the energy produced by nuclear reaction, but no radiation is produced.

It is estimated that 10 mg of antimatter (the size of a grain of salt) can produce the thrust equivalent to that produced by 200 tons of liquid chemical fuel, which can send a huge rocket into space at the rate of one-third of the speed of light—at this speed, it will take only 2 years to fly through the solar system. One gram of antimatter can make a car run for 100,000 years. Yet, it is very difficult to develop and produce antimatter; production cost is extremely prohibitive. It has been arbitrarily estimated that to produce 1 g of antimatter, we will need to spend at least $1 billion. Besides, it is difficult to store and transport antimatter because it will explode as soon as it touches common matter.

Although countries around the world are all relentlessly exploring antimatter, our study of it is just at the primary stage, far from practical use. We hope it will be used by people someday and become an epoch-making smart energy.

Geomagnetism technology. Geomagnetism (the earth's magnetic field) can be considered the magnetic field of an imaginary magnetic dipole situated at the earth's center. Its south and north poles are opposite to the geographical South and North Poles. Its south pole roughly points to the geographical North Pole. Its north pole roughly points to the geographical South Pole. The magnetic field of the earth's surface changes because of many factors. The weakest magnetic field is near the equator (about 0.3–.094 oersted) and the strongest is at the two poles (about 0.7 oersted). There are many theories to explain how geomagnetism formed.

> The first theory was that there is a giant magnetic iron core, which makes the earth a huge magnet. This theory was proved wrong because even though the earth's core is full of iron, nickel, and other matters, their magnetic properties are lost as the temperature increases to up to 760°C—the temperature of the earth core is as high as 5000–6000°C; so it is impossible for these materials to serve as the huge magnet in the earth.
>
> The second theory is that geomagnetism is produced from the earth's ring current. Because the temperature at the center of the earth is very high, iron, nickel, and other matter are in the molten state; they rotate with the rotation of the earth, making the electrons and charged particles

have directional movement. In this way, ring current is formed and produces geomagnetism like a charged helix tube. But, this theory cannot explain the several reversions of geomagnetism that have occurred in history.

The third theory is that geomagnetism is the result of the mutual action of the conductive fluid inside the earth and the magnetic field inside the earth. In other words, there is a magnetic field inside the earth. The rotation of the earth drives metal matters to rotate, because of which the induced current is produced. The induced current produces an external magnetic field. This theory is also called the "geodynamo theory." The premise of the theory is that there is a magnet inside the earth, but the theory cannot explain its source.

Although the reason why geomagnetism is produced needs authoritative scientific explanation, the existence of geomagnetism is an indisputable fact. The earth itself is a ceaselessly rotating and revolving moving body with huge kinetic and geomagnetic energy. If we could make enough closing coils to cut geomagnetic lines, the energy produced by this would be tremendous, taking us to a whole new future.

18.2 System Framework of Smart Energy

A system is a common set of rules we must abide by under certain conditions; its presentation form is usually laws and policies. Technology and system cannot exist in isolation. A certain technology must be coordinated with a certain system to promote its development and make it be fully used by society. And certain systems must also be gradually shaped, improved, and developed based on certain technologies. A smart energy system is formulated to overcome the disadvantages of the traditional energy system. It not only has its own complete and strict system but is also linked, naturally integrated, undividable with, and even difficult to distinguish from other human systems because smart energy itself will be involved in many aspects of our life.

The smart energy system is involved in energy development, production, processing, storage, transportation, conversion, consumption, recycling and global cooperation; it is closely related to everyday life. It not only takes laws and policies as expressed form but also includes value belief, ethical norm, moral concepts, customs, ideology, etc., and finally rises to the comprehensive system of energy politics, economy, culture and other aspects. It also includes the system of promotion of energy saving, environment production, and cooperation.

In the course of our civilization evolution, energy technology and energy system came from the same source but had different names; they intertwined and affected each other. At the primary stage, energy technology was dominant, and the energy system auxiliary. As social activities and energy technology became increasingly complicated, the energy system became increasingly important. Since modern times began, the status of the energy system is as important as the energy technology, and in some aspects, even more important. The human intelligence contained in the energy system has been equal to the energy technology, played the same role, achieved the same goal as it, and promoted its development, for example, an energy system can also make energy cleaner and more efficient. Because the intelligent nature of the energy system and the energy technology is increasingly evident, their roles and purposes are increasingly similar and undividable. The energy system and energy technology finally will be combined into smart energy and become inseparable. Their combining level embodies intelligence. Smart energy is the epitome of highest intelligence, and its technology and system are completely combined.

In the hunting–gathering period of civilization, energy supply could completely satisfy our demand, and the effect of firewood energy on the natural environment was so little that it could be neglected; energy development, use, production, and consumption were completely spontaneous and voluntary; and the energy system was perhaps not as important as today and in the future—the task assignment of simply collecting firewood and social division of labor was the energy system in its embryonic form. In the agricultural period of civilization, our energy use forms were richer and more complicated; energy system began to be shaped—the ownership of cattle, horses and other domestic animals, the investment and benefit of water wheels and wind wheels, and so on were relatively well regulated. In the periods of the industrial civilization and information civilization, technologies rapidly changed and emerged in an endless stream: coal and petroleum and other energy are consumed in large amounts, nature has been overburdened, and increasingly dwindling energy resources and imbalanced geographical distribution of the same results in innumerable disputes. It is especially important to determine energy ownership, production, development, investment, benefit and consumption, and other aspects through a proper system. The system was gradually developed and improved, and now begins to be integrated with technology in intelligent attributes. In the future period of ecological civilization, the smart energy system will play a more important role and be completely integrated with inseparable from smart energy technology.

In the future, energy systems will contain more and more human intelligence and complement energy technology. With regard to intelligence attributes, the technology and the system of smart energy will gradually change from imbalance and separation to balance, coordination, and integration as an inseparable whole, producing smart energy as the new energy form. The smart energy system framework in the transient period mainly includes the following four aspects (see Figure 18.2).

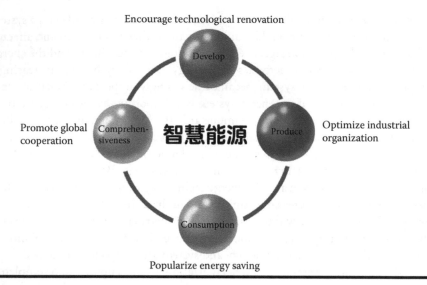

Figure 18.2   Smart energy system framework.

Encouraging scientific innovation. The fundamental purpose of the smart energy
   system is to promote the innovation of smart energy technology, and its basic
   characteristic in design is to encourage and realize innovation. Because smart
   energy technology needs to overcome the huge obstacle in cost before it
   becomes mature and can be massively utilized, the smart energy system must
   fully encourage the spread and use of clean and high efficiency innovative
   technologies and alternative technologies.

Optimizing industrial organization. An important mission of the smart energy
   system is that through a whole set of systems, the industrial organization of
   energy should be promoted to be more efficient, conform to industrial object
   law, obtain maximum output with minimum cost, and realize scale benefits
   to ensure that energy supply can meet the demand of economic and social
   development.

Advocating energy saving. The smart energy system pays attention to the creation
   of an advanced consumption philosophy that will change all social members'
   consumption behavior, rebuild new energy saving fashion, and encourage
   them to use energy saving products and clean products and carry out energy
   saving and emission reduction with real action.

Promoting international cooperation. As economic globalization develops in depth
   and the interdependency of world countries is increasingly strengthened today,
   the concept of sharing benefits and responsibilities is more and more suitable
   for energy research, development, production, consumption, environment production, and other fields. The smart system will fully promote (Figure 18.3)

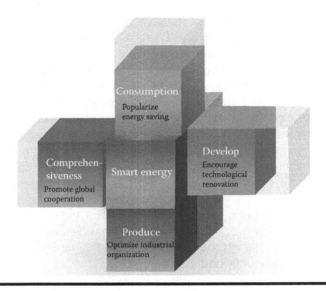

Figure 18.3   The institutional framework of smart energy.

and strengthen international cooperation to integrate isolated and scattered forces to convert them into the composite force of the whole mankind and promote our successful transformation toward the future civilization.

Smart Energy System Rises in the World

In smart energy, technology is "hardware," and its system is "software." The two complement each other. A system without technology as its foundation is just an illusion, and technology without system as guarantee is limited only to laboratories. Smart energy needs the mutual promotion of technology and system. Smart Energy Technology embodies the Smart Energy System; the Smart Energy System facilitates Smart Energy technology advancement. At present, the smart energy system has risen in the world. All kinds of energy systems are constantly emerging ranging from the cooperation and exchange between states and organizations to innovation and exploration between industries and enterprise, full of strong vitality.

Energy Performance Contracting (EPC). EPC is a market-based brand-new energy saving mechanism and energy management method gradually developed in market economy countries since the middle of the 1970s. It has developed faster and become involved in more energy saving fields in the U.S. and Canada in the past 10 years, because the government gave importance to it, capital was sufficient, and Energy Service Company (ESCO) credit system was relatively perfectly established. The essence of EPC is the energy saving investment method that uses reduced energy saving cost to pay all the investment of energy saving projects, allow consumers to use future energy saving benefits to carry out energy saving projects, or allow energy saving service companies to provide energy saving services to consumers by promising the energy saving benefit of energy saving projects or by contracting all energy costs. The energy saving services contract is signed by and between the enterprises (users) that carry out energy saving projects and the energy saving service companies. It helps to promote the implementation of energy saving projects that are technologically feasible and economically reasonable. EPC can be divided into sharing EPC, promising EPC, and energy cost trusteeship EPC.

EPC came to China in 1997. China's National Development and Reform Commission, World Bank, and Global Environment Facility jointly developed and carried out the "Work Bank/ Global Environment Facility China Energy Saving Promotion Project" and established model EPC companies in Beijing, Liaoning, and Shandong in China. The internal rates of return of these companies' projects have been over 30% for many years after establishment. China has accumulated lots of EPC experiences through over 10 years' development. In China, EPC projects have played a vital role in energy saving, especially in building energy saving, green lighting, electric motor system energy saving, waste heat and pressure use, and other fields. But they themselves have the characteristic of high risk and long cycle. It is necessary to intensify policies, perfect industrial regulations, and strengthen spread and propagation.[*]

Demand Side Management (DSM). DSM involves management activities to improve terminal electricity use efficiency and change electricity consumption modes with the support of governmental regulations and policies, taking effective encouragement and instructive measures and suitable operation modes and through the concerted efforts of power generating companies, grid companies, energy service companies, social intermediary organizations, product suppliers, electricity consumers, to reduce electricity consumption and demand while satisfying the same electricity demand function to achieve

[*] See *Study on EPC Projects Commercial Operation Modes and Scheme*, Li Hongdong, Post-doctoral final report, Chinese Academy of Sciences, November 2012.

resource savings, environmental protection, and maximum social benefit, thereby benefiting all parties and minimizing cost energy services. The U.S., European countries, and other developed countries all have a great number of people who are engaged in DSM work. In 2000, the U.S. put about $1.56 billion in DSM, saving 53.7 billion kWh and reducing peak load by 22 million kw.

In the early 1990s, the power companies and electricity users in China conducted a great amount of electric DSM work under the direction of the government, such as increasing peak electricity price, implementing interruptible load electricity price, and other measures to instruct users to adjust production and operation mode, adopt ice-storage air-conditioning system, and heat storage electric boilers. At the same time, some encouraging policies and measures were adopted to spread energy saving lamps, variable frequency electric motors and pumps, high-efficiency transformers, and other energy saving equipment. It is estimated that during the 11th Five Year Plan, through implementing electric DSM, China saved over 100 billion kWh electricity and over 60 million tons of coal. If effective electric DSM continues to be carried out, by 2020, China can reduce installed capacity by 100 billion kW, exceeding the installed capacity of five Three-Gorge Projects while saving power investment of 800–1000 billion RMB, which not only will greatly relieve resources, environment, and investment pressure but also will bring about huge energy savings, environmental, economic, and social benefit.

Energy finance. It realizes continuous optimization and combination of the energy industry capital and finance capital through integration of energy resources and capital resources to promote a series of financial activities of benign interaction and coordinated development of the energy industry and the finance industry. Because of the special status of energy and finance in economic development, energy finance is both a strategic problem in energy and finance development and a core problem in economic development.

The government can set up an energy finance system through energy regulators, financial regulators, energy enterprises, banks, investment funds, etc., and encourage and support energy enterprises or financial institutions to establish energy strategy banks to issue securities to promote energy resources and capital circulation and appreciation. At present, the countries of the world are very interested in the development of energy finance, and their energy finance markets continuously pursue and play games around petroleum prices to some extent. The contract value of the financial derivatives trading that BP, Shell, and other transnational energy enterprises participate in each year reaches hundreds of billions of dollars, and trading objects include contracts of futures, options, forward and interchange, etc., of commodity markets, interests markets, and foreign exchange markets.

The energy finance in China started relatively late but has developed rapidly. At present, the energy industry and the finance industry have been combined, and energy securities, energy funds, energy futures, and other financial products have been derived. The earliest financial derivative product was petroleum futures. Shanghai Petroleum Exchange launched petroleum futures in 1993, which exceeded the Singapore International Monetary Exchange one year later and became the world's third biggest energy futures trading market. The energy section has been closely linked with national energy policies and energy industry performances and is flexibly interactive. China's low-carbon industry fund, green energy development fund, and other energy funds have rapidly developed, widening the financing channel for the energy market.

A Long Way Ahead: How Do We Get There?

Perpetual power is the dream we pursue. Smart energy enables our dream to come true and give us a future full of hope. But the exploration of smart energy still has a long way to go, especially in the innovation of energy development, the use technology, and reform of the energy production and consumption system. Finally, to ensure success on our road to the final destination, we must induce the dawn of smart energy into a red sun that shines brightly all over the world so as to let future civilizations thrive in full sunlight and power.

Chapter 19

Global Mission: A Very Long Way to Go

19.1 The Globe Is Simultaneously Cold and Hot

In accordance with the law of conservation of energy, energy neither emerges out of the void nor disappears without foundation. It is only converted from one form to another or is transferred from one object to another; its total quantity never changes in the course of conversion and transfer. Our use of energy is to transfer energy substantially from stable carrier media, such as petroleum and coal, to our living space through combustion and other methods. Besides, due to the comprehensive action of greenhouse gases, the total energy input in our living space is more than the total energy output from our living space, which makes accumulated energy increase and all kinds of media become more and more active. Air containing reinforced thermal energy and kinetic energy will be more likely to form hurricanes, land containing excessive heat will intensify water evaporation resulting in desertification, and rivers containing excessive heat will dry up. These factors will lead to ecological environment damage and climate change.

Based on the law of conservation of mass, in chemical reactions, the mass sum of all substances before a reaction is equal to the mass sum of all the substances generated after the reaction, and the total mass of any system isolated from the surroundings never change no matter what change or course has happened. Most raw materials release energy through chemical reaction in the course of energy use. In the left side of the chemical equation, the substances before the reaction, namely raw material, will become less, while in the right side of the chemical equation, the substances generated after the reaction, namely waste or pollutants, will become more. It can be deduced from this that the direct result of energy development and use is energy depletion and environment pollution and the global warning against which we have been warned incessantly.

We have mastered the law of conservation of energy and the law of conservation of mass but we are far from mastering the law of balance of ecological environment. At present, when the world population and economy are growing rapidly, minerals, land, fresh water, forests, wild animals and plants, and other natural resources have gradually become depleted. Global climate warming, ozone layer depletion, acid rain, water resource worsening, soil resource degradation, global forest crisis, biodiversity reduction, poisonous and harmful substance pollution, and transboundary movement and other ecological environment problems can threaten mankind at any time and at any place.

No country, region, nation, organization, or individual is excluded from globalization. We all need to face energy, ecology, environment, and climate problems because we are under the same blue sky and on the same earth.

The countries of the world should proactively determine their responsibility in accordance with the historical total consumption of energy and resources and damage of ecological environment to protect our common earth. First, we must promote the clean use of energy and concentrate our efforts on pressing problems, such as climate change and environment pollution. Second, we make long-term efforts to resolve the power problem met in the course of human development and promote research and use of smart energy. Developed countries not only should be responsible for the historical debt of their high-carbon development since the Industrial Revolution but should also give technical and financial support to developing countries for energy saving and environmental protection. Developing countries should

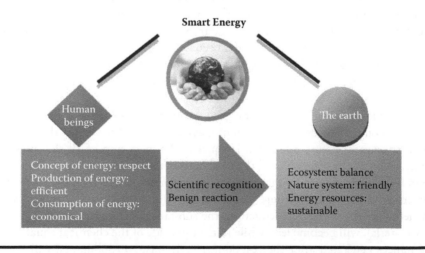

Figure 19.1 Harmonious energy concept in ecological civilization—mankind and the earth shake hands.

make scientific plans to avoid taking the old path followed by developed countries of controlling events after the pollution has taken place; instead, they should take early initiatives for energy saving, environment protection, and ecological restoration.

Smart energy has established an unprecedented bridge and link between us and the earth. We should give up avarice, impudence, and ignorance and set up a respectful, highly efficient, and thoughtful approach and realize ecological balance and sustainability of energy and resources through scientific cognition, benign interaction, repair, and restoration. It is only in this way that we can build a more friendly and harmonious relationship with the earth—let the earth become green, healthy, and full of vitality and let our lives become more comfortable, sunnier, and happier (see Figure 19.1).

Earth Hour

Earth Hour is an initiative proposed by the World Wide Fund For Nature for coping with global climate change. Under this initiative, individuals, communities, enterprises, and governments turn off lights for one hour (20:30–21:30) on the last Saturday of March each year to show their support to the action of coping with climate change. At present, the climate change resulting from excessive carbon dioxide emission has greatly threatened human survival. The threat on the world can be alleviated only by changing people's attitude toward carbon dioxide emission.

Earth Hour started in Sydney, Australia on March 31, 2007, and rapidly swept across the world. The theme of Earth Hour 2012 was "there is an environmentalist in everyone's heart," and the purpose was to encourage individuals, enterprises, and governments to make a change in environment protection at least. On that day, a lot of cities, enterprises, and individuals voluntarily participated in Earth Hour and turned off unnecessary lights for an hour. The activity inspired people to think: In addition to turning off lights, can we do more?

Everyone's small step together is one big stride for Earth. The purpose of Earth Hour is not merely for saving electricity but for making the public recognize the importance of protecting the earth and improving their awareness of environment protection—to encourage the good habit of environment protection in daily life. Earth Hour proposes the following suggestions:

> *Green transportation.* Choosing public transportation as the primary means of mobilization; unplugging electric appliances after use; using energy saving windows and doors with better heat insulating capacity; drying clothing in the

open air; using shopping bags instead of one-off plastic bags; setting air conditioning appliances to a suitable temperature in order to avoid too big a difference between indoor and outdoor temperature; turning on the water fountain only when you need to drink water; storing food not more than 80% of the fridge volume.

Rubbish recycling. Rubbish is treasure in the wrong place. Help to dispose of rubbish from sources reasonably; use separate garbage cans at home for kitchen garbage and other waste; collect ring-pull cans, pop bottles, etc., to sell them to waste recyclers; collect rainwater for cleaning and watering flowers.

Forest protection. Carry out paperless office work; use electronic greeting cards instead of paper; refuse to accept leaflets, folding, or other forms of advertisements; read electronic magazines and get news online; use electronic bills; refuse to use disposable chopsticks; buy furniture with FSC certification*; celebrate festivals and commemoration days by planting trees.

Fresh water conservation. Using the same water that we used for washing our face to wash our feet, watering flowers with water used for washing rice, mopping the floor with the water used for washing clothes; closing the faucet while brushing teeth and immediately after use; using water-saving close stools and faucets; washing vehicles with recycled water or vegetable wax; taking cups with you when you go out and attend conferences; using phosphor-free detergent; not throwing away waste into rivers and lakes; concentrated recycling of discarded and old batteries.

Wild animal protection. Do not discard waste at will, as this will prevent wild animals from eating it; do not hunt or fish outdoors; appreciate wild animals and plants, do not touch or pick them up; do not eat wild animals, especially endangered wild animals; do not buy ivory and tiger skin; do not wear animal furs as they are not fashionable but cruel; do not pick wild flowers and grass while traveling; do not buy wild animals as pets, let them live with their own families.

---

* The global forest problem has become more and more prominent. Forest area is reducing and forest deterioration increasing. In 1990, some American consumers, timber trade organizations, environment and human right organizations believed that it was necessary to create an honest and creditable organization to identify forests run conservatively and accept them as sources of forest products. For this, the Forest Stewardship Council (FSC) was formed.

LOHAS*. Buy vegetables and fruits in season produced locally to avoid excessive transportation cost; choose organic fruit and follow a vegetarian diet as much as possible; do not smoke; support social charities, donate old articles; reduce garbage, classify and reduce garbage; ride vehicles less and take more exercise; use air conditioning less, commune with nature more; hold a unique low-carbon wedding; buy products with green environment protection marks, etc.

19.2 A Long Way Ahead

Common but different principles must be adhered to in order to promote smart energy development. Developed countries must take the main responsibility because this is determined by history and reality. Developed countries are the earliest and greatest troublemakers of climate warming and environment damage and have unshakable historical responsibilities. Developed countries are also economically and technologically dominant, have greater strength and abilities, and so also have inescapable realistic responsibilities. Developing countries contribute two-thirds of the present total emission reduction, but the developed countries have resisted change in their development pace, unwilling to reach the promised target. Climate, environment, and resource problems have tired us beyond endurance, and strains in global cooperation have made our future grimmer.

Developed countries accounted for 95% of the total carbon dioxide released by fossil fuel combustion from the 18th century (when Industrial Revolution began to 1950) and 75% of the global total carbon emission in the subsequent 50 years when they accounted for only 15% of the global total population. That the developed countries have emitted huge amount of carbon dioxide and other greenhouse gases without restraint and used resources indiscriminately in their course of industrialization and modernization is the main reason for the present global warming and environment damage.

Developed countries have moved their industries with high emission or low-end manufacturing links to developing countries by virtue of the high monopoly of technologies and standards in the present global division of labor in manufacturing process. Their per capita carbon emission has far exceeded that of developing countries since the Industrial Revolution.

The developed countries that have earlier finished the course of industrialization and urbanization have advanced technologies, deep pockets, and comprehensive strength for promoting smart energy development. It is evident that they have

---

* LOHAS, Lifestyles of Health and Sustainability.

greater responsibilities—provide necessary capital and technical support to developing countries, make due contribution to the improvement of climate warming and environment damage, and make efforts for the exploration of smart energy so that the whole humanity can tide over difficulties and crises smoothly to enter the future civilization together. This is not only the developed countries' responsibilities but also conforms to their and the whole world's long-term interests. But some of these countries do not regard these responsibilities of coping with the global crisis, which is embodied in the following aspects:

- Developed countries refuse to implement emission reduction on the ground that developing countries fail to participate in them and that it affects economic development. The U.S. population accounts for 3%–4% of the global total population, its carbon dioxide emission is equivalent to over 25% of the global total carbon dioxide emission, but it proposes that China, India, and other developing countries assume the same emission reduction indexes, or it will not take any responsibility of emission reduction. The European Union always persists that it is meaningless to only have developed countries conduct emission reduction; its emission reduction is far less than the increase of emission of some developing countries. The U.S. government announced that it would refuse to ratify the Kyoto Protocol in March 2001. Canada declared that it would withdraw from the Kyoto Protocol in December 2011 just after the Durban Convention ended, refusing to carry out its responsibilities of emission reduction.
- Developed countries delay financial support on the ground that the developing countries must intensify emission reduction and accept international supervision. The United Nations Framework Convention on Climate Change (UNFCCC) passed in 1992 states that developed countries shall support developing countries in finance and technology, but the reality is not so rosy. Developed countries vowed at the World Climate Conference in Copenhagen to provide 30 billion dollars during the rapid starting period between 2010 and 2012 and 100 billion dollars by 2020. But they failed to implement this. Besides, at present, the aid capital that the developed countries promised to provide to developing countries for coping with climate change and that has been carried out by the developed countries is just $2 per capita in the developing countries.
- Developed countries postpone technical transfer on the ground of protecting intellectual property rights and private ownership of property rights. The developing countries' energy-saving and emission-reducing technologies are relatively backward, and lots of key energy-saving and emission-reducing technologies of the six high-consumption and high-emission industries of power, building, transportation, metallurgy, cement, and chemical engineering were controlled by developed countries. They should have kept the promise they made during the Convention and provided timely and sophisticated technical aid to the developing countries. Regrettably, many developed countries worry that the international competitive power of their domestic

industries and products will be affected after the transfer of their advanced technologies, and therefore they are unwilling to transfer these powers, citing taxation and administration and other means as the ground for the protection of intellectual property rights and for the private ownership of many technologies. They pay one-sided emphasis to exerting the role of private sectors and market and condemn the developing countries stating that the domestic system and environment are not helpful to technical transfer, resulting in little results in technical transfer for more than 20 years.

The fact that developed countries shirk their responsibilities is perhaps helpful for their own interests in the short term but harmful for the whole world including them. This is the famous "frog-boiling effect" in biology. The frog will leap out of the boiler with boiling water and fall on the ground safe and sound, if it is immediately put into it. But if the boiler is filled with cold water when it is put into it, and it is allowed to swim freely, while the water is heated with a small fire, though it feels the temperature change, it does so gradually and because of inertia it becomes too numb to escape when it cannot endure the heat. The greenhouse effect is like heating the earth with a small fire. Though there is no dramatic change in a short time, the consequence of long time accumulation is destructive. The developed countries are like the frog that is gradually heated. They now do not make efforts to reduce emission and even want to increase emission. But the whole world's economic interests, including theirs, will come to naught when global warming is irreversible. The consequence will be faced by both them and the whole world.

These reasons or pretexts come from the developed countries' economic ethics. The majority of the developed countries base their economic ethics on capitalist private ownership, under which market basis manufacturers with clear property rights and consumers, buyers, and sellers, the two parties, both pursue maximized interest, no matter whether a micro economy or state is so. A country that gives consideration to its own national interest is blamed for no reason. However, if it only gives consideration to its own national interest and ignores other countries' interests even at the cost of harming other countries' interests, in reality, it is harming its own interest from a long-term point of view.

There is no national boundary in climate warming and environment damage. No country can shirk responsibility, and the developed countries can go faster and farther on the road of developing smart energy and coping with global crises if only they seek common ground while resolving differences and making concerted efforts with the developing countries.

Doha Conference: A Show of "Pride and Prejudice"

The American delegation wasted valuable time when technical transfer and other essential problems should have been discussed in the team negotiation process at the first stage of the 18th UNFCCC Conference held in Doha, Qatar, in 2012.

Canadian delegates claimed, "we are here for negotiation, not commitment" when they were asked whether they would provide capital to the green climate fund. Japanese delegates announced the total carbon emission of the countries bound by the Kyoto Protocol (KP) only accounted for 26% of the global total carbon emission to justify that their refusing to sign the second commitment period was "reasonable." Furthermore, many developed countries still tried to impose unreasonable emission reduction indexes on China, India, and other developing countries.

Some developed countries have unfulfilled responsibilities in emission reduction commitment and set obstacles in technical aid in a show of "pride and prejudice." By pride, we mean they have financial, technical, and industrial advantages and can therefore surpass the bearing abilities of developing countries. By prejudice, we mean that they evaded duties by a variety of methods and even made attempts to shift emission reduction pressure on developing countries and let them lose their acceleration on the path of development. But, developed countries should examine themselves and set an example for others to follow, whether in history or in reality.

As can be seen from history, climate change is mainly caused by the greenhouse gases emitted by developed countries for a long period of time; this is the consensus of the international community. It is estimated that about 80% of the present greenhouse gases in the atmosphere belong to developed countries' historical emission; the top countries in per capita emission are mainly developed countries. In other words, the developed countries are the main perpetrators of climate change, and the developing countries are the main victims. However, so far, the perpetrators still shirk their responsibilities for a variety of reasons and even attempt to shift responsibilities to the victims.

In reality, the developed countries' per capita GDP is several times higher than that of the developing countries; some developed countries' greenhouse emission is still increasing. But the developing countries face the arduous mission of eliminating poverty, developing economy, and improving people's livelihood while accepting large amounts of transfer emission because they are in the low end of the international industrial chain. The developed countries ignore reality and require the developing countries to bear responsibilities beyond their abilities.

If the goal of controlling global mean temperature rise to within 2°C is achieved, the greenhouse gas emission from developed countries will be reduced 25%–40% in 2020 from that in 1990. But their commitment is only about 15% on average; furthermore, whether they stick to their commitment or not is debatable. Among them, America only makes a commitment of emission reduction of 17% in 2020 from that in 2005, equivalent to an emission reduction of 3% compared to the value in 1990. In contrast, some developing countries have taken ambitious emission reduction action. For example, the Chinese government has made a commitment that by 2020, unit GDP carbon emission reduction will be 40%–50% that of 2005, which fully shows its good faith and that it took the revolutionary approach of coping with climate change. In this important subject that is related to the future of the whole mankind, the developed countries should first give up their "pride and prejudice" and private interests and show more responsibility, awareness and sincerity, or they can only widen the economical gap between people and hinder negotiations on climate change—a bitter pill that will have to be swallowed by the whole of mankind.*

19.3 General Mobilization of Smart Energy

Smart energy construction is a new competition in the course of human civilization evolution. Different from the past, this is a team competition, with the whole mankind as contestants and nature as the referee. If we break the rules, we, the whole human race, will be completely eliminated. With solidarity and cooperation, we can achieve great common victory!

The winner of the Nobel Prize in Economics, George Arthur Akerlof,[†] said "Control of global warming is not a problem of cost and benefit but a problem of morality." The development of smart energy needs global intelligence, and the whole mankind will share the result of it. What we face are common difficulties and hopes. The crisis in nature will not wait for us to finish quarreling and passing the buck. This is not a problem of who takes responsibility but a series of problems of how many responsibilities a country should take, which responsibilities it should take, and how it takes its due responsibilities to the end. We should more deeply experience the urgency of reaching a consensus and quickening action after we have already sustained huge losses brought on by endless debates. The developed countries must base themselves on reality, look forward to the future, and make concerted efforts with the developing countries. The developing countries should also take action and make their own voices heard to safeguard the earth's safety together.

Objectively, although some developed countries have been evading their responsibilities using a variety of reasons for a long time, they have really made an indelible contribution in the development of smart energy and coping with global crisis and are making their own efforts. Having consumed a large amount of energy and undergone all kinds of throes brought on by the industrial civilization, many men of insight in the developed countries, even at relatively earlier times, began to reflect on the limitations of the industrial civilization starting from the actual national conditions—they recognized the importance and necessity of the development of smart energy. They proposed a related civilization development direction and the advanced concepts of the ecological civilization, including green GDP, environment justice principle, ecological ethics, environment sociology, sustainable development, circular economy, etc. At present, these concepts have been generally

---

* See Doha Climate Negotiation—Refuse "Pride and Prejudice," Li Zhihui, Yang Yuangyong, Chen Ying, Xinhua News Agency, December 3, 2012.
† George Arthur Akerlof, American economist, professor of economics of the American University of California at Berkeley, won the Nobel Prize in Economics with Michael Spence and Joseph Stiglitz in 2001.

recognized and become the principle needed to be jointly agreed on for world development. International energy organizations and other institutions cannot do without the developed countries' cooperation and efforts—the International Energy Agency and Organization of Petroleum Exporting Countries were all established under the promotion of the developed countries.

The developed countries have jointly developed all kinds of smart energy technologies—hydrogen energy, combustible ice, solar energy, wind energy, and geothermal energy, to name a few—using their financial and technical advantages and through governmental introduction and enterprise as main bodies. They first developed marine energy, nuclear energy, shale gas, and other new technologies of environment protection. Accordingly, they have enacted and perfected all kinds of systems to restrain the use of high-consumption, high-pollution energy forms; formulated resource mining standards and the supporting financial backing, investment, price, and taxation policies; and tried for the systematic creation of smart energy. But, all these are just the base; developed countries still need to consciously take and carry out more responsibilities and make more efforts in technology, system, and cooperation.

This does not mean that developing countries should look on with folded arms and do nothing in coping with global crisis, even if they are restrained by backward technology. As mentioned earlier, their contribution has been nearly two-thirds in present total emission reduction. They accept low-end manufacturing links with high consumption, high pollution, and high emission, and in fact even bear the developed countries' transfer of carbon dioxide emission and resources consumption. Even so, China, India, Brazil, Mexico, and other developing countries still spare no efforts to take proactive action in coping with climate change and adopt a series of policies and measures for controlling greenhouse gas emission and coping with climate change, such as adjustment of economic structure, improvement of energy structure and energy use efficiency, and development of sustainable energy and bringing this into the national plan.

Developed countries should provide more technical and financial aid to the developing countries in the future. Smart energy needs innovative technologies. The developing countries are not advanced in technical innovation, and the developed countries must provide technical aid to them free of charge—they cannot take it as a means of facilitating their own national capital and commodities output and seeking expansion of influences and effect in aid-receiving countries. The technical aid must be advanced technology, not backward and outdated technology. Convenient policies and systems, must be provided for technical aid, and the obstacles to protection of autonomous intellectual property rights must be eliminated. In financial aid, the developed countries should strengthen economic cooperation with the developing countries based on equality and mutual benefit and help them in capital management and use. The results of the smart energy should belong to the whole mankind—the developed countries provide technical and financial aid to the developing ones, and the developing countries will provide market and employment opportunities in turn when they use smart energy technology. In this way, a global benign interaction cycle can be achieved to drive us toward a higher ecological civilization.

Smart energy is the new dawn of human civilization. We live in hope even if it still does not radiate brilliant light. We should maintain sufficient confidence and enough patience. Development and use of smart energy not only can ease energy crisis and pressure but can also meet the demand of the present and future development of human civilization. It is also the power that promotes civilization evolution and accelerates civilization transformation. Once the concept of smart energy is ingrained in the hearts of people and once the development of it is like a spark causing a prairie fire, civilization evolution will obtain inexhaustible power, green earth and ecological civilization will be combined perfectly, environment pollution will be controlled and will gradually become a thing of the past, we will no longer worry about environment deterioration, and the countries of the world will no longer fight for energy shortage and energy safety.

Global Initiative

Promotion of smart energy to benefit the world is right, a general positive trend, everyone's duty, and the unshakable responsibility of the great powers. The developed countries have already completed their course of industrialization and urbanization and have advanced technology, abundant financial power, strongest instructing and driving force, and comprehensive advantage in promoting smart energy development, which is their major responsibility and glorious mission as well as the great trust and hope the developing countries had for the developed countries.

The exploration of new smart energy technology should be targeted. The whole of mankind can share the development results of smart energy only when the developed countries spearhead the work of development, spread, and use of smart energy. Their development approaches to smart energy should be diversified—they not only can do many things at once and work together to overcome technical and systematic difficulties in a short time but also can proceed in an orderly way and work independently to exert their respective comparative advantages to the hilt and overcome difficulties by themselves.

The promotion and improvement of the smart energy system should keep pace with the times. Efforts should be made to promote the continuous development of the smart energy system to make it adapt to smart energy technology and even promote its progress. On one hand, effective, detailed, and perfect systems should be formulated in smart energy research, development, production, processing, storage, transportation, conversion, consumption, recycling, and cooperation to guarantee and protect smart energy development. On the other hand, energy production and consumption conceptions should be changed to create a good social environment and systematic basis for smooth research, development, spread, and use of smart energy.

The promotion of international cooperation in smart energy should join hands. Smart energy matters and is related to long-term strategy. There is no one country or region that can succeed in it easily. Hence, international exchange and cooperation in smart energy technology and systems must be proactively promoted. The developed countries should take full consideration of the developing countries' development stages and basic demands, respect their appeals, and closely combine the development of smart energy, coping with climate change, thereby promoting the developing countries' development and improving of their sustainable development abilities.

At present, crisis breaks out; difficulties are clustered; and resources, environment, climate and other problems threaten human survival. We have a long way to go for a new civilization, and ecological civilization calls for new energy power. The developed countries should keep the concept of "peace, development, and cooperation," look to the world and the future, worry ahead of other people, and enjoy the fruits of labor after them. They must not be short-sighted and mutually pass the buck; they must bear more responsibilities, conform to global common interests, and leave the blue sky, fertile soil, and clean rivers to future generations. We fully believe and sincerely hope that the brilliant sunshine of smart energy will shine and benefit the whole world under the leadership of developed countries and by global coordination.

Chapter 20

China's Responsibility: Rationally Taken

20.1 China's Conscious Action

During the 30 years of reform and opening up to the outside world, China as a developing country has gradually entered the middle and last stage of industrialization, upgraded consumption structure, accelerated urbanization, and made remarkable economic achievements, but it has also paid an expensive price. In 2003, the Chinese government put forward the Scientific Outlook on Development to build an energy saving and environment friendly society and began to promote conscious action toward smart energy.

On the energy saving and consumption reduction front. In industrial boilers, cogeneration, and other fields, 10 key energy saving projects and 1000 energy saving actions for enterprises are being implemented, the energy saving management of key energy consuming enterprises is being strengthened, and energy audit and energy efficiency benchmarking activities are being promoted. In the manufacturing technological field, green design technology, new processing technology of energy saving and environment protection, green disassembly, and recycling and remaking technologies are being established to promote energy saving and consumption reduction in the course of industrial production and product use. In building energy saving, the implementation rate of the compulsory energy saving standards for newly built buildings is being raised, the energy saving transformation of existing buildings is sped up, the scale application of sustainable energy in architecture is promoted, and the energy saving transformation of office areas in public institutions is being carried out. In the transportation field, fuel consumption limits and an access system of operating vehicles are being implemented, special action of low-carbon transportation of thousand cars, ships, roads and ports enterprises is taken, and urban public transport is vigorously developed.

Energy cleaning and use. The proportion of coal washing and dressing has been increased, and coal transportation and direct combustion and use are reduced. Using high quality coal to generate electricity has been encouraged. Integrated gasification combined cycle, large supercritical circulating fluidized bed, ultra-supercritical power generation units, and other clean power generation model engineering constructions are proactively pressed ahead to raise the clean coal power generation ratio. Development of catalyzer series products for engineering application is encouraged, and the first commercial operation of direct coal liquefaction projects was conducted. A combination of water power development and ecological environment protection is considered, the activities of environment protection work of projects that have been built and are being built is emphasized, the development and use of water power environment protection technology is intensified, and green water power assessment standards and systems are enacted. Measures of reinforcement of power development and management, improvement of the coordination of wind power and grid and support for the development of wind power equipment enterprises, etc., create conditions for massive development and use of wind power. Solar thermal power generation model engineering pilot projects have been established in Inner Mongolia, Gansu, Qinghai, Xinjiang, Tibet, and other suitable regions to stably promote the development of the solar power industry. The safe management of nuclear power projects that are being built is implemented well, and the safe and stable operation of the nuclear power generating units that are operating is guaranteed.

Environmental protection. Pollution control systems and standard systems were established and have been perfected to strengthen environmental pollution control. Environmental pollution is examined in an all-round way, including assessment of pollution sources and the production, emission, and pollution control of the main pollutants. Emission reduction by 10% of the two main pollutants—sulfur dioxide and chemical oxygen demand—is taken as the binding indexes of national economic and social development, and total pollutant control is vigorously promoted by adopting comprehensive measures, such as engineering emission reduction, structural emission reduction, and management emission reduction. Laws and regulations are being gradually established and perfected to regulate the management of hazardous medical and electronic waste. Environment management legislation for chemicals is proactively promoted to strictly implement import and export environment management registration of new and poisonous chemicals.

Ecological protection and restoration. Ecological civilization construction and ecological protection and construction engineering are carried out in accordance with the principle of the harmonious interaction between man and nature. National Main Functional Zone Planning clearly put forward the goal and mission of building "two screens and three zones" ecological safety strategy pattern in order to protect the ecological environment. The Revolution to Quicken Forest

Development has been launched to establish a forest development strategy with ecological construction playing a dominant role. The prairie protection and construction engineering projects, such as returning grazing land to grasslands, grassland harnessing of the karst regions in southwest China, and nomad settlements, are being carried out. China Action Plan for Wetland Protection provides action guidance for wetland protection, management, and sustainable use. China Action Plan for Biodiversity Conservation and China Compendium of Natural Reserves Development Plan (1996–2010), agricultural and forestry plans and other industrial plans are being issued and implemented, and a series of actions for biodiversity protection are taken. The natural reserves network that is relatively reasonable in layout, relatively complete in nature, and relatively sound in functions has been formed; the ex situ conservation of wild animals and plants and germplasm resources are being rapidly developed. Investigations based on biodiversity, scientific research, and monitoring ability are being greatly improved, and biological safety management is being strengthened.

Global climate change. The system and mechanism of coping with climate change are being gradually established and perfected in accordance with the related provisions of The United Nations Framework Convention on Climate Change and KP and are combined with the general requirement of sustainable development strategy. The China National Assessment Report on Climate Change was first issued in 2006; China National Plan for Coping with Climate Change was published and carried out in 2007, making clear the guiding ideology, main fields, and key tasks of coping with climate change; the Twelfth Five Year Plan for Controlling Greenhouse Gases Emission was enacted and issued in 2011, giving all-round directions for the control of greenhouse gas emission; and the Second China National Assessment Report on Climate Change was issued in the same year. Coping with climate change will continue to be brought into economic and social development plans and will continue to adopt powerful measures in the future: (1) to strengthen energy saving and strive for obvious reduction of unit GDP carbon dioxide emission by 2020 compared to the emission in 2005; (2) to vigorously develop renewable energy and nuclear energy and fight for nonfossil energy consumption ratio reaching about 15% of primary energy by 2020; (3) to increase forest carbon sink and forest area increase by 40 million hectares by 2020 compared with 2005 and forest storage increase of 1.3 billion m^3 compared with 2005; (4) to develop green economy, low carbon economy, and recycling economy and develop and spread climate friendly technology.*

Not only the Chinese government but also Chinese enterprises including state-owned and private enterprises are all taking conscientious action and throwing themselves into the development of smart energy. During the Eleventh Five Year Plan, the Aluminum Corporation of China's energy consumption per 10,000 yuan added value reduced by 26%, unit energy consumption of aluminum oxide reduced

---

* See Chinese President Hu Jintao's speech Join Hands to Cope with Climate Change Challenge given at the Opening Ceremony of the UN Summit on Climate Change on September 22, 2009.

by 36%, comprehensive energy consumption of aluminum unit product and copper unit product respectively reduced by 23% and 33%, and chemical oxide demand emission reduced by 66.63%. Devotion Corporation has successfully developed biomass fuel and its liquefying and gasifying technologies, replaced traditional fossil fuel with them, used them in boilers and other industrial fields, and obtained good economic benefit. Biomass fuel is ecological, zero emission, renewable, and obviously has social benefit. ENN Energy Holdings Limited proactively developed a universal energy machine using universal energy grid technologies, conducted supply and demand conversion and matched, gradient utilization, space and time optimization of all kinds of energy flows through the coupling of energy production, storage, transportation, application, and recycling with information so as to maximize system energy efficiency and finally outputs a kind of self-organizing highly orderly energy.

Scientific Development Leads to Smart Energy

The Eighteenth National Congress of the Communist Party of China was very clear on the importance of China's future development. The road to the all-round, coordinated, sustainable ecological civilization and scientific development it depicts must promote the rapid development of Chinese smart energy. The following aspects are its main characteristics:

Expansion of new energy form. Saving all kinds of resources, especially energy, is the fundamental strategy of protecting the ecological environment. National energy safety is guaranteed by the promotion of an energy production and consumption revolution, control of total energy consumption, reinforcement of energy saving and consumption reduction, support for the development of energy-saving low-carbon industries, new energy, and renewable energy. Vigorous support for the development of new energy and renewable energy will effectively promote the rapid development of smart energy.

Promotion of energy technology progress. A new economic development approach should be rapidly formed in accordance with the new changes in domestic and overseas economic situation so as to shift from the standpoint of promoting development to improving quality and benefit. Efforts should be made to stir up a new vitality in development of all kinds of market bodies, reinforce the new power of innovative development, and establish a new system of modern industrial development. Technical progress should be driven by relying more on modern service industry, strategic newly emerging industry, saving energy, and recycling economy. Intelligence should be taken as an important content of the "new four modernizations." The new four modernizations require "continuing on the road to progress with the Chinese characteristics of new industrialization, informationization, urbanization, and agricultural modernization, promoting the deep integration of informationization and industrialization, the benign interaction of industrialization and urbanization, the mutual coordination of urbanization and agricultural modernization, and promoting the synchronous development of industrialization, informationization, urbanization and agricultural modernization." The scientific innovation and technical progress that the new four modernizations advocate must promote smart energy technology.

Promotion of energy system innovation. Energy system innovation should be intensified, binding ecological environment examination indexes and economic environment indexes should be carried out, and pilot projects of energy saving, carbon emission, emission, and water rights trading should be conducted. Energy saving trading means the market trading behavior that all kinds of energy using units (or government) indulge in by buying and selling energy saving (or energy consumption right) under specific energy saving goals and in accordance with the goal-finishing situation. Carbon emission right trading means that government institutes assess the greatest carbon dioxide emissions that meet environment capacity in a certain region, divide it into carbon emission rights and puts them into the market. Emission right trading means internal pollution sources mutually adjust emission through current exchange under the premise that total emission in a certain region is not more than the allowed emission for that region. The three trials of market-oriented reform all belong to the smart energy system. It can be seen that the energy system innovation represented by these reforms will greatly promote, enrich, and perfect the smart energy system.

20.2 China's Rational Response

China has realized rapid and stable economic development, remarkable improvement of people's living standards, and made much progress in the control of total population, improvement of population quality, resources saving, and environment protection, etc. At the same time, as a populous developing country, China is ecologically weak—less per capita resources, prominent regional development imbalance problem, and poor scientific innovation; China works very hard for improving livelihood, increasingly bound by resources and environment in economic development, and still has 122 million poor people. There exist difficulties and hopes, and opportunities and challenges. China needs to rationally treat the current predicament as development opportunities brought by smart energy.

Rationally coping with all kinds of challenges. The present climate warming, environment pollution, resource shortage, energy disputes, and other challenges are directly related to unreasonable energy development. China, which is populous and territorially vast, cannot take a development road that only relies on traditional energy, but should proactively promote smart energy development and press ahead with its own development. The world's present minable energy has been unable to satisfy the demand of China, India, and other countries realizing industrialization. China realizing modernization also cannot be at the expensive cost of polluting Mother Earth. China can resolve the current contradictions only if it shoulders the lofty historical mission of making great contributions to human society, rationally coping with challenges, and proactively developing smart energy.

Coping with all kinds of challenges caused by energy development and use is the unshakable common responsibility of the whole human society. As an important member of the world family, China should bear important responsibilities, which is an inevitable requirement for China's own development. However, China is globally the biggest and most populous developing country and is very imbalanced in development. It must, on the basis of its national condition and ability, seek truth from facts, rationally cope, act according to its ability, and persist in its own development baseline. All countries and nations have equal rights to subsistence and development. China represents itself and extensive developing countries; it has the moral responsibility to bear what it should bear, and not bear what it need not bear.

Reasonable treatment of the development of smart energy. China should rationally understand the important meaning of smart energy, pay enough attention to it, and reasonably treat its development.

1. The pace of developing smart energy should be moderate. China cannot go all out and fast like a swarm of bees in the development of smart energy. This will waste resources and cause safety problems. Thinking about the road China has already taken to smart energy, we easily find its development is rapid. New energy industry is listed as a strategic new industrial field of key development and is supported by making a series of laws, regulations, and policies. By the end of 2011, Chinese water power, wind power, nuclear power, liquid biomass fuel, and other nonfossil energy production had accounted for 8.3% of that year's total primary energy consumption. At present, China has four firsts in new energy development, namely, it is the first in water power installed capacity, the first in solar heater use scale, the first in nuclear

power construction scale, and the first in wind power installed capacity. But, the rapid development of new energy has also brought two major problems. One is the quality safety problem that is becoming increasingly prominent. Because wind power, photovoltaic energy, and other new energies take on the characteristics of randomness, intermittence, and volatility, the massive connection of new energy into the grid may produce harmonic waves, adverse current, too high grid voltage, and other problems, posing a challenge to the safe and stable operation of the grid. The other problem is the partially excess production capacity. In terms of the present industrial situation, some of the strategic new industries, such as wind power, photovoltaic energy, seriously exceed production capacity. The Central Economic Working Conference held in December 2012 proposed that "control over excess production capacity would have an effect on wind power and photovoltaic energy." How to regard and control the excess production capacity of these industries is an important problem at present.

2. The dimension of developing smart energy should be suitable. Dimension here means technical approach, namely, what way is adopted to promote the development of smart energy technology. The development approaches of smart energy can be roughly divided into two kinds—"energy development" and "energy saving." The former means developing new energy, such as promotion of the innovation of new energy technology and transformation of new energy system. The latter means the innovation of traditional energy technology, including energy saving and emission reducing technology and system. Of course, China should carry out energy development and energy saving strategies in developing smart energy, but it should pay more attention to energy saving and energy saving technology and systems in the near future. This is because China is one of the few countries that mainly consume coal. Its primary energy production totaled 3.18 billion tons of standard coal in 2011, ranking first in the world. Developed countries basically consume petroleum and natural gas. China has great potential in reducing its coal consumption. This is also because Chinese per capita energy consumption is relatively low; in 2011, it was only 2.59 tons of standard coal—just about the world average level, and far lower than that of America, Japan, South Korea, and other countries. Chinese people do not use energy like American people. So energy saving is of more realistic meaning in the near future in Chinese smart energy development.

China's Rights and Duties in Emission Reduction

In the implementation of carbon dioxide emission reduction duties, developed countries of the western world always have

their own ideas—they weaken and evade the equality principle prescribed in KP 1997 and the "common but different
responsibility" principle with extensive international consensus
and hope that the developing countries including China shoulder relatively high, clear, and verifiable emission reduction
tasks, but fail to provide the kinds of technologies and finance
they had promised to provide to developing countries; besides,
they fail to fully carry out the emission reduction duties they
should implement. At the Durban Climate Summit held in
November 2011, the Chinese delegation put forward the notion
that UNFCCC and KP be legally binding international conven-
tions and that all parties should abide by and keep their own
commitments. Five questions must be resolved before a new
global emission reduction protocol is reached. They all are the
stipulations that have been determined by international climate
negotiation and should be carried out.

1. The first commitment period of all countries' emission
   reduction stipulated by KP, the first international legal document restraining greenhouse gases emission in human
   history, ended in 2012, and an international legal protocol
   that regulates the second commitment period for emission
   reduction of all countries, Paris Agreement, is made to
   ensure the realization of the global goal that "temperature
   rise is controlled within 2°C at the end of the century."
2. Developed countries pay a rapid launching capital of
   $30 billion and a long-term capital of $100 billion per
   year by 2020 and start the green climate fund as fast as
   they can. Enforcement supervision mechanism should be
   established for provision of capital and technical transfer.
3. The consensus reached in technical transfer, forest restoration, transparency, and capacity construction should be
   carried out; and related mechanism should be established.
4. The assessment of each country's implementation of
   its commitment should be stepped up and issue the report
   for controlling the temperature rising up to 1.5°C at the
   year 2018.
5. The principle of "common but different responsibilities"
   should be adhered to, and it should be determined in
   accordance with the wholeness and history of environment protection; each country should take the responsibilities and duties that are suitable for its own development
   stage and level.

20.3 China's Dream of Smart Energy

Smart energy is a brand new energy that accelerates civilization transformation and upgrading and meets the demand of future civilization. It is related to China's major plan of improving national competitive force, building a beautiful country, and realizing national rejuvenation. China missed opportunities in the first and second Industrial Revolutions. Since the establishment of the People's Republic of China, especially since the reform and opening up of the country, the energy industry has been sustainably and rapidly developed, which has laid the foundation for a good future in smart energy; China has made theoretical preparations for smart energy and accumulated valuable experience regarding this. But not only China, all world countries have marched toward smart energy and launched the new round of energy revolution. Under the pattern of economic globalization and social informationiza-tion, China must lose no time to grasp the strategic opportunity of smart energy. It must proactively speed up its own development, initiatively integrate itself into international competition, go all out to occupy the first position in smart energy technology to drive the all-round, healthy, and sustainable development of the soci-ety and economy, and help to promote and lead human civilization evolution. The country that has smart energy has the strategic ability to dominate global energy and can reach the top of the global energy system before other countries. China, which is territorially vast, populous, and has undergone rapid economic development and has immense energy demand, needs to achieve important innovation and progress in key energy technologies and systems. China must rely on its own huge market and scientific research and develop the potential to become the leading country in smart energy development and play a dominant role in the hard-won transformation. Smart energy will promote the strategic restructuring of Chinese economic structure. The realization of the energy saving and emission goal of 2020 and longer term develop-ment is the reliable guarantee of realizing China's energy safety, cleanness, and high efficiency, building a world-leading new energy structure, embodying scientific com-petitiveness, and setting up a new international energy standard and order.

At present, Chinese smart energy is in the early and exploratory stage; the intelligent elements contained in energy are few, but there is a good development momentum. Chinese economy and society are in the postindustrial period; tradi-tional energy technology still dominates; the evils and defects of the double-track market system are relatively obvious; and problems, such as resources restraint, environment pollution, and ecological deterioration are severe. Based on the pres-ent situation, China should adopt a short-term, safe, and rapid approach to break through technical bottlenecks, eliminate systematic obstacles, and address pressing problems. Looking into the future, China should proceed from technology and system in the development of smart energy, make top designs and strategic layout, encourage scientific innovation, optimize industrial organizations, advocate energy saving, promote international cooperation, and set up a systematic, safe, clean, and economic smart energy system.

Decoding smart energy technology. To develop smart energy technology, we should first strengthen and improve traditional energy technology; vigorously support the highly efficient clean use of coal, high-efficient clean conversion of heavy oil, and solar energy conversion; develop and use nonconventional petroleum resources; combine cycle power generation of distributed energy system and distributed gas turbine; improve, research and expand water power generation, wind power generation, fuel cells, hydrogen energy, fuel for clean cars, and other technologies to make sure of achieving real change in the near future and resolving presently prominent contradictions while laying a solid foundation for the transformation and upgrading to the next stage. To develop smart energy technology, we still need to make the strategic blueprints of new energy technology—encourage scientific research, invention, and creation; explore and develop cold energy, solar wind, antimatter, geomagnetism, human body energy, and other technologies; expand compressed air, flying wheels, and other energy-storing technologies; broaden superconduction, wireless transmission, and other transmission technologies; promote intelligent grid, universal energy network, and other comprehensively using technologies; and launch pilot engineering in time to gradually form a productive force. Besides, new technologies in fields such as cloud computation, energy satellites, artificial intelligence, nanotechnology, genetics, biogenetics, and ecological environment protection should be promoted to provide support services for smart energy technology.

Creating smart energy system. Smart energy needs an effective scientific system as guarantee. The smart energy system can be divided into a scientific research system, production system, and consumption system in accordance with the energy industry chain. In scientific research, universities, scientific research institutes, enterprises, and related organizations should be encouraged and supported to devote time and effort to innovation and research of smart energy. In production, industrial structure should be optimized; backward, high-consumption industries should be eliminated in time; and highly efficient, clean, low-carbon new technology should be promoted and used. In consumption, consumers should be instructed and encouraged to uphold thrift, stop waste, and practice smart use of resources. The smart energy system can be divided into a national and international systems in accordance with geographical scope. Nationally, the laws, regulations, and policies related to smart energy should be established to guarantee that the related activities of the energy system are regulated and orderly. Internationally, exchange and cooperation should be further strengthened to carry on the exchange of needed goods, learn from one another's strong points to make up one's deficiencies, and realize a win–win situation.

The development of smart energy in China can be divided into three steps: within 20 years, the energy structure should be changed from focusing on fossil energy to focusing on clean energy, an internationally leading innovative smart energy technology system should be established, the key layout of alternative smart energy technology should be established, and breakthroughs should be made; within 50 years, alternative smart energy technology should be basically perfected, matured, and applied, gradually become a major energy developing and using technology, and should play an important role in international energy; within 100 years, smart energy should become the main power of civilization evolution, replace traditional energy in an all-round manner, and help China dominate global energy development, finish the transformation, and upgrade toward ecological civilization while promoting economic and social sustainable development.

A Letter to Home

Human history is very long; the future of the universe is limitless, and everyone is just a fleeting passerby. We are fortunate enough to live in such a great epoch full of imagination, exploration, change, and innovation, so much so that there is new hope and objective each time the sun rises; there is valuable gnosis and gain each time night falls. Stars in the sky are picturesque and they enlighten human wisdom.

Chinese civilization stretches 10,000 years; Chinese people have undergone the change of vast sea becoming farmland. Chinese history has gone through many dynasties from Da Yu harnessing floods to Xia, Shang, Zhou Dynasties, Qin uniting six states, Chu and Han states contending against each other, the rise and fall of the two Han Dynasties, the flourishing Sui and Tang Dynasties, the prosperity of the two Song Dynasties, the Mongolian Yuan rivalry, the rising of the Ming Dynasty, and the decline of the Qing Dynasty. Years are like the change of seasons. The Chinese nation suffered greatly from disasters and calamities in the Opium War, the Eight-Power Allied Forces invasion, and Japanese aggression. China also experienced Hundred Days' Reform, the Revolution of 1911, the Northern Expedition, the 8 years of war against Japanese aggression, liberation from imperialism, feudalism and capitalism, and the founding of a new China. It has now entered a period of openness and reform and is in the process of national rejuvenation.

The rise and fall of the nation is the concern of every citizen! China has always taken the lead and flourished; the nation has been defeated many times and many homes have been lost; it lost its way and missed opportunities many times. The past cannot be recalled, and we cannot start again. But we cannot look down upon ourselves, nor can we be overbearing. If every Chinese person starts introspecting, any tiny thought and exploration, multiplied by the 1.4 billion, leads to unparalleled great wisdom and power. If everyone is ambitious, united, and advances bravely, any difficulty and challenge, divided by

1.4 billion, is like a praying mantis trying to stop a chariot—not worth mentioning. Each citizen should consciously save resources, protect the environment, offer advice and suggestions, and proactively take action; each family should support and help each other to build beautiful communities together; each enterprise should bravely shoulder heavy burdens and implement the duties of energy saving and emission reduction to protect green mountains and rivers; and scientific planning and overall development should be carried out in economic construction to realize the harmonious development of economy and ecology.

The world tide goes forward with great strength and vigor; those who go along with it live, and those who go against it die. Nations rise and fall following the natural law—the survival of the fittest. Historical opportunity is fleeting and will not come again. Smart energy is China's opportunity; China is the stage of smart energy. China has chosen smart energy, and smart energy has also chosen China. To build a moderately prosperous society and a beautiful China and realize the nation's rejuvenation, we need smart energy. China's cardinal number of population, economic scale, and development prospects are the natural incubator for smart energy and provide favorable scope for its use. China's wisdom should meet with the world's and give out heat and light on the same earth and under the same blue sky. China's smart energy should implement benign interaction, mutual benefit, a win–win situation.

Any road that leads to success is filled with both flowers and brambles. To reap where one has not sown is a pipe dream. To press on and work hard can benefit people and help the nation prosper. Chinese people have an undefeatable spirit, seething enthusiasm, and limitless wisdom. Under their persistent exploration and unremitting efforts, China's smart energy must make great progress and become the inexhaustible power for promoting Chinese social and economic development and civilization evolution. China must make great contributions to the worldwide development of smart energy.

Conclusion: Create Our Future

Our future is filled with countless unknowns and possibilities. Although it cannot be accurately predicted, the possibility of any event depends on the choice we make. The journey of 10,000 years has shown us that we can certainly overcome difficulties and obstacles and promote civilization evolution as long as we continuously conform to the demand of civilization evolution, make unremitting efforts, and proactively explore new energy forms. We firmly believe that human beings, as the smart creatures in the universe, must adopt smart energy technologies to create a brilliant future civilization.

Energy Gives Us Fundamental Guarantee

Energy provides the fundamental condition necessary for the evolution of civilization. No creature can do without energy if it wants to live. In addition to subsistence, the evolution of human civilization also depends on energy.

It has taken mankind a very long time to learn and use energy and promote civilization evolution. To break the tether and look for more freedom is the highest pursuit of human civilization. The discovery of fire at a certain point in the past was a momentous occasion. It was an unspeakably wonderful experience, turning as from chaos towards wisdom, taking a step forward owing to mankind's exploring nature and using natural force to get rid of restraint and gain freedom and liberation.

With fire, we got rid of starvation, acquired adequate food to live and multiply and created the hunting–gathering civilization, which was not enough. With wind power, hydropower, and animal power, we eliminated the obstacle of rivers and obtained the freedom of migration and expansion and created agricultural civilization, but even this was not enough. With the steam engine driven by coal, we created the industrial civilization, cast off the fetters of gravitation, and obtained the freedom to fly in the sky and explore the depths of the ocean, which was also not enough. With coal, petroleum, and electricity, we have created the wonderful contemporary epoch of information civilization, gotten rid of atmospheric obstacles, and acquired the freedom of chasing after the stars and the moon, but even this is not enough. We are eager for more freedom.

From fire to water wheels, windmills, carriages, and sailing boats and from the steam engine to the internal combustion engine and power generators, we got enough food to eat, sailed on rivers and seas, traveled around the world, and explored stars. Through the intelligence accumulated through many generations and civilizations and through technical exploration, mankind is capable of changing energy forms, improving energy efficiency, developing and using new energy, and pushing human civilization and freedom to a higher level. Without the silent support of energy, our civilization and freedom will be like water without a source and like a tree without roots. Our civilization achievements change rapidly and our energy power transforms heaven and earth, but even this is not enough. What freedom and power will our next civilization and next generation of energy bring to us?

History Tells Us Direction of the Future

Energy replacement is the objective law of civilization evolution. Xunzi said, "a gentleman is not different in nature but good at working with materials." In the long history of mankind, we have continuously used the great power of energy, changed and developed the forms of utilizing it, ceaselessly improved the efficiency, and reduced its negative effect on resources and environment, and so have survived, developed, and written brilliant chapters in the evolution of civilization.

Ten thousand years ago, our ancestors began to control natural forces with their wisdom of drilling wood to make a fire. It was much more difficult than domesticating wild animals into domestic animals and wild plants into crops and vegetables. It was bumpy and even heart-stirring. But, the more tortuous the road is, the more it can inspire human intelligence and enthusiasm. We tried to steer sailing ships using the wild and intractable wind and thereby found new and fertile lands to inhabit. We tried to make water wheels to steer the flow of water to irrigate the land and feed future generations.

The invention of the steam engine greatly enhanced productivity and guided people to enter the new era of industrial civilization. Coal, which was called a "black monster" by Marco Polo, helped people drive trains and oceangoing ships, which helped to facilitate trade to different places in the world. Moreover, the emergence of the internal combustion engine showed human intelligence and petroleum, known as industrial blood, was flowing in cars, planes, and submarines, and thus human civilization cut a conspicuous figure. The discovery of electromagnetic induction and appearance of power generators lifted the curtain to the electric era, started the second wave of industrialization. From then on, science and technology have changed the way we live.

The industrial civilization greatly enriched human fortune and deeply affected the natural environment. The large amount of greenhouse gases released by fossil fuel combustion have led to global warming, iceberg melting, rising sea levels, and some island countries facing catastrophe and some inland countries suffering industrial pollution. Blue sky and white clouds have become wished for in the cities, stinking and polluted streams surround villages, crystal clear rivers with frolicking fish and shrimp have become a distant memory. Under the concept of "whoever controls petroleum resources controls the whole world," wars and petroleum are tightly bound to each other, resulting in frequent wars; avarice and hegemony have pushed industrial civilization into a bottomless chasm.

The historical wheel is rolling forward and is carrying mankind from the hunting–gathering civilization, agricultural civilization, and the industrial civilization to the information civilization. At this moment, if we look back at the past, we can clearly see that the improvement and replacement of energy forms have always enabled civilization evolution. Today, if fire had not been discovered and put to use, mankind would probably still be savage and would still be short of food if natural forces had not been domesticated, the industrial revolution would have perhaps just been a pipe dream without the use of coal, petroleum, and natural gas, and the prosperity of the contemporary information civilization would have been just science fiction without the appearance of electric energy.

Our ancestors discovered many brilliant things, and the benefits of these are seen even today, and thus they have handed us the baton of civilization evolution. Bravely exploring new energy and injecting a more powerful driving force into the historical wheel is the sacred mission that we must undertake.

Wisdom Endows Us with Great Power

Smart energy is the inevitable demand of civilization evolution. Intelligence forms the footstone for civilization and promotes energy replacement. Without intelligence, mankind is no different than birds and animals. Energy is difficult to be replaced. Different forms of energy have different requirements. The higher the civilization level is, the higher the requirement of energy forms will be, the more human intelligence contained in the way human consume the energy. Energy can meet the requirement of the continuous development of human civilization only if it is continuously produced and developed. Our civilization developed from the hunting–gathering civilization, agricultural civilization, and industrial civilization to the present-day information civilization; energy forms have gone from initial smart energy, lesser smart energy, and medium smart energy to greater smart energy; future ecological civilization needs the strong support of real smart energy, which is the inevitable requirement for civilization evolution. The ecological civilization uses smart energy, and smart energy is characteristic of the ecological civilization. The two are like engaging gears to drive human civilization to go toward a more brilliant future step by step. So, how will smart energy be born?

Smart energy is a mix of human scientific and technical innovation, which is the fundamental power of exploring smart energy and core content of it. In the history of civilization evolution, so many dreams regarded as impossible were realized like magic because of the technical innovation driven by human intelligence, and the various technologies from the drilling of wood to make a fire to the contemporary information technology are all engraved with human intelligence and are deeply changing the world and promoting the progress of civilization. Smart energy solidifies human system changes. It is difficult to effectively use new energy technology in an all-round way without an effective system. If technical innovation is the spark, system changes set the prairie ablaze. Mankind must maximally promote the faster evolution of human civilization through more intelligent system arrangements. Smart energy strengthens human cooperation efforts. International cooperation provides the strong support necessary for promoting the development of smart energy. A single hand can't clap. In the vast universe, we independently live and rest and only rely on ourselves. Only when the whole of mankind unites, can we advance triumphantly on the future road of human civilization to solve our problems related to energy, environment, and climate change.

There are flowers and thorns on the road of smart energy and ecological civilization just like the production of the important raw material polycrystalline silicon is accompanied by the generation of the hazardous chemical silicon tetrachloride. The soil it touches turns barren. The turbines of wind power generators can easily cause harm to migratory birds, thereby affecting the whole ecological chain. There are many bottlenecks in science and technology that are hard to break through, and system construction is also not perfect.

Although thorns block our way and the future world is still a mystery, we are still driven to make unremitting efforts to explore smart energy. Let us convert thorns into roses. There will be exuberant ecological civilization, clean smart energy, vigorous scientific innovation and system reform, and green seas and blue sky brought about by international cooperation in the future. With the dream, we will be more steadfast and persistent in the journey to the future, our civilization will ceaselessly continue, and we can find inexhaustible energy power full of intelligence in historical changes.

In the past 10,000 years, our ancestors left many wonders and surprises, much wealth and joy to us, and energy, which has evolved from fire to electricity. In addition to giving them much thanks, we should carefully think of what we can leave to our future generations. Let us create the energy technology legend that can be left to our future generations. Freedom belongs to intelligent mankind and glory belongs to smart energy, and we forever cherish most ardent longing for a better future.

Mankind never stops pursuing for dreams!

Postscript

The "we/us" in the book refers to all the people on earth, including the readers, and the "we/us" in the postscript refers to three people: Liu Jianping, Chen Shaoqiang, Liu Tao, respectively, who were born in the 1960s, 1970s, and 1980s. We have different life experiences, backgrounds, and occupations, but we have common concerns and common thoughts for the future.

Liu Jianping was born beside the Yangtze River in the countryside beside Dongting Lake. His childhood memory is still clear and beautiful. There was luxuriant green grass and brightly colored flowers in spring; he played in the water in rivers and lakes in summer and even fished and picked lotuses; there were fruits in autumn, and the land was covered with snow in winter. These images often appeared in his dreams even after he grew up. He has many feelings towards time and space, one of which is that it took a whole day to go to town and back 40 years ago, while now it takes the same amount of time to go to Beijing and back. Forty years ago, keeping in touch with relatives and friends generally took 10 days or half a month and was mostly through letters, but now people can stay in touch at any time and from wherever they are. This is the progress and comfort that have been brought about by development, he and his contemporaries can see and enjoy. Although there are many such happy experiences, there are also many unspeakable anxieties and worries. The streets in cities are widened, buildings are taller, there are more cars, standard of life is improved, but air, soil, rivers, lakes, and even seas are polluted; resources are in short supply; good faith is rare. The people under the same system certainly have the same behavior. If he had not made it to middle age and had not played the social roles with great significance, he would have never had the time to think and would not have thought of the problems discussed in this book.

Chen Shaoqiang was born in Xiaogan, Hubei. His hometown is famous for sesame seed candies and filial piety. Different from his father's generation, he did not suffer from hunger in childhood but had a hard life. In the course of his growing up, he lived in the countryside and then moved to cities and even to foreign countries, and the word he most came into contact with was "double-track system".

The most lasting impression he has of those years is of the rapid changes that took place. Buildings rose straight from the ground one by one and fridges, TVs, and mobile phones that were once considered luxuries rapidly became mass consumer goods. Now he needs the internet every day, and the bicycle is still his favorite means of transportation.

Liu Tao was born in a rural family in the region along the middle and lower reaches of the Yangtze River and loves the brightly colored, idyllic scenery and the simple honest rural customs. Unfortunately, these now only exist in his memory. The river where he usually went to catch tadpoles in his childhood has dried up, there have been many factories on that piece of land where golden rape were grown, childhood companions have subsequently moved to cities and have since married. Fortunately, happening to meet Mr. Liu and Mr. Chen, he became one of the three authors of the book.

We even completed some projects in industry, regional development, and financial policies. With further job changes and subsequent professional field expansion, we gradually began to be involved in ecological environment, energy and resources, system, and other related areas. In the course of studying them, we all obviously felt the restraints of our respective professional fields and the shortage of comprehensive knowledge. The professional self-conceit caused by the confidence in the fields we were engaged in made us ignore or disrespect other fields in the past. So, we increasingly tend to rethink the gains and losses in our work in recent years.

The Chinese President Hu Jintao once said at the APEC CEO Summit held at Busan, South Korea, in November 2005, "in the history of human social development, each important progress of human civilization is accompanied by energy improvement and replacement," which was impressive. We always thought of studying the issue but failed to do so because of time constraints, and only at the end of 2009, we (in the attitude of learning, improvement, and challenging ourselves) joined forces and studied the civilization and energy issues, which was also many people's concern.

For over three years, the book was written intermittently because there was no research related to this issue nor funds available, and because we were all very busy and had many things to attend to. Besides, we had troubles that we were reluctant to discuss or mention, because with the deepening of our writing, we found that there were too many disciplines to cover and two many subjects to study, which were far beyond our knowledge and abilities. But we did not give up. We continuously adjusted the outline and even revised its title a few times while ceaselessly learning. However, we never changed the subject and plotline, namely civilization evolution and energy replacement.

For the overarching questions of energy and environment, science and ethics, and history and future, we shortened the long history, reduced professional jargon to plain language, supplemented abstruse science and technology with pictures, and lightened the serious facts as much as possible. Although we worked very hard and carefully, we were very clear that some of our standpoints were just the statements of one school and that many questions still need to be explored further. No matter how much the market economy tide surges, we as three individuals among about 7 billion people on earth only want to express our concerns, thoughts, and responsibilities. We believe there are always people, both in the past and even in the future, concerned about this issue to different degrees—we are sure we three are not the only ones.

There is still another important reason for our not giving up on the book: the encouragement and support from all circles of life. We thank the leaders and colleagues of The State Electricity Regulatory Commission (hereinafter referred to as the SERC): Wang Qiang, who is both strict and lenient, whose help and recognition provided us the impulse to finally finish the book; erudite yet modest Sun Yaowei (a good teacher), who always gave us valuable advice; persistent and shrewd Yang Mingzhou who continuously assessed the book; Wu Jiang, who is good at both literature and sports, who carefully listed a series of books we needed; and Ren Lixin, who is specializing in science and technology, could always help us find the suitable words to describe the scientific facts and terminology. Jiang Zhaoli of the Department of Climate Change of National Development and Reform Commission; Li Fulong and He Yongjian, leaders and experts of National Energy Administration; and other leaders and experts who provided us with enthusiastic guidance. Zhao Jingzhu and Wu Gang (the two research fellows) and Qian Tiejun, Zheng Xiancao, XiongXiangfu, Jiang Hui, and Li Chunming (the five doctors of the Institute of Urban Environment, Chinese Academy of Sciences) imparted much wisdom to us. Xu Hong and Tan Zhenyong (two doctors of the State Grid Corporation of China), Zhu Jinping (Chief accountant Hebei Electric Power Corporation), Qi Xiaoyao (President of the Shaanxi Regional Electric Power Group Co., Ltd.), Yuan De (Chief Engineer of the China Power Investment Corporation), Ying Guangwei (President of the China Huadian Corporation, China Electric Power Research Institute), Jia Bin (a doctor and research fellow of ENN Energy Holdings Limited), Wu Kehe (Director of Beijing Engineering Research Center of Electric Information Technology, North China Electric Power University), Qu Changfu of the Economic Daily, and various other friends also lent us support. Wu Shaojie of the Southern Branch of China Huaneng Group, Yang Jian of Hubei Education Examinations Authority, and Tan Tian of School of Information Science and Technology, Tsinghua University, did most of the basic work for the book. The book also has referred to and quoted the results of many domestic and overseas research studies. It is the very crystallization of these experts' wisdom that provided the basis for our writings. We sincerely thank all the abovementioned and unmentioned leaders, experts, and friends. We deeply regret any errors in this book, and we would be happy if you could correct them for us.

We especially thank Cheng Siwei, Vice President of the Standing Committee of the Tenth National People's Congress, for writing the preface for the book and Yang Kun, former chief engineer of State Electricity Regulatory Committee, for his examining the first draft of the book.

Bibliography

Anderson, C. *Makers: The New Industrial Revolution*, Chinese Edition. Translated by X. Xiao. Beijing: China CITIC Press, 2012.

Botkin, D.B. and D. Perez. *Powering the Future: A Scientist's Guide to Energy Independence*, Chinese Edition. Translated by Caomu. Beijing: Publishing House of Electronics Industry, 2012.

Bryson, B. *A Short History of Nearly Everything*, Chinese Edition. Translated by W. Yan, et al. Beijing: Jieli Publishing House, 2005.

Cao, R. *Global Warming: Economics, Politics and Moral of Climate Change* (In Chinese). Beijing: Social Sciences Academic Press, 2010.

Central Compilation & Translation Bureau. *Karl Marx and Frederick Engels, Volume 3* of *Selected Works of Marx, Engels, Lenin, and Stalin* (In Chinese). Beijing: People's Publishing House, 1995.

Cheng, D. *Introduction to the Top Design of Smart City* (In Chinese). Beijing: Science Press, 2012.

China Institute of Contemporary International Relations. *Global Energy Structure* (In Chinese). Beijing: Current Affairs Press, 2010.

China Telecom Smart City Research Group. *The Path of Smart City: Scientific Management and City Personality* (In Chinese). Beijing: Publishing House of Electronics Industry, 2012.

Chinese Academy of Sciences Energy Area Strategy Research Group. *China's Energy Science and Technology Development Roadmap* (In Chinese). Beijing: Science Press, 2009.

Climate Change Science and Technology Policy Research Group. *Overview of Climate Change Science & Technology Policy in Major Developed Countries and International Organizations* (In Chinese). Beijing: Science and Technology Documentation Press, 2012.

Crosby, A.W. *Children of the Sun: A History of Humanity's Unappeasable Appetite for Energy*, Chinese Edition. Beijing: China Youth Press, 2009.

Encyclopedia of Energy Editing Committee. *Encyclopedia of Energy* (In Chinese). Beijing: Encyclopedia of China Publishing House, 1997.

Engdahl, F.W. *A Century of War: Anglo-American Oil Politics and the New World Order*, Chinese Edition. Beijing: Intellectual Property Publishing House Co., Ltd, 2008.

Ferguson, N. *Civilization: The West and the Rest*, Chinese Edition. Translated by X. Zeng, et al. Beijing: China CITIC Press, 2011.

Freese, B. *Coal: A Human History,* Chinese Edition. Beijing: China CITIC Press, 2005.

Geller, H. *Energy Revolution: Policies for a Sustainable Future*, Chinese Edition. China Environmental Science Press, 2006.

Intergovernmental Panel on Climate Change (IPCC). *Special Report on Emissions Scenarios.* IPCC, Geneva, 2001.

International Energy Agency. *Energy Technology Perspectives: Scenarios & Strategies to 2050*, OECD/IEA.

International Energy Agency. *Energy Technology Perspectives: Scenarios and Strategies to 2050*, Chinese Edition. Beijing: Tsinghua University Press, 2009.

Jin, C. et al. *Big Power's Responsibility: China's Perspective* (In Chinese). Beijing: China Renmin University Press, 2011.

Jin, X. *The New Energy Development Report of China* (In Chinese). Wuhan: Huazhong University of Science and Technology Press, 2011.

Krupp, F. et al. *Earth: The Sequel: The Race to Reinvent Energy and Stop Global Warming*, Chinese Edition. M. Chen, et al. translated. Beijing: The Oriental Press, 2010.

Letcher, T.M. *Future Energy: Improved, Sustainable and Clean Options for Our Planet*, Chinese Edition. Translated by T. Pan. Beijing: China Machine Press, 2011.

Li, C. *New and Renewable Energy Technologies* (In Chinese). Nanjing: Southeast University Press, 2005.

Li, X. et al. *Power: 50 Scientific Theorems That Have Changed Human Civilization* (In Chinese). Shanghai: Shanghai Culture Publishing House, 2005.

Lin, B.Q. *Fiscal Policies to Promote the Development of Renewable Energy Research* (In Chinese). Beijing: China Taxation Publishing House, 2010.

Lin, B.Q. and G.X. Huang. *Energy Finance* (In Chinese). Beijing: Tsinghua University Press, 2011.

Liu, J. *Towards a Higher Civilization: Multidimensional Perspectives for the Exploitation of Hydropower Resources* (In Chinese). Beijing: People's Publishing House, 2008.

Liu, Z. *Smart Grid Technology* (In Chinese). Beijing: China Electric Power Press, 2010.

Lovins, A. *Reinventing Fire*. Chelsea Green Publishing, 2011.

Lynch, Z. et al. *The Neuro Revolution: How Brain Science Is Changing Our World*, Chinese Edition. Translated by Y. Bao. Beijing: Science Press, 2006.

Martin, N. et al. *Emerging Energy-Efficient Industrial Technologies*, Lawrence Berkeley National Laboratory/American Council for an Energy-Efficient Economy. Report No. LBNL-46990, Berkeley, CA/Washington, DC, 2000.

McCully, P. Discredited strategy, *The Guardian*, https://www.theguardian.com/environment/2008/may/21/environment.carbontrading, May 20, 2008.

Melosi, M.V., *Waste*. Beijing: Foreign Language Teaching & Research Press, 2004.

Montgomery, S.L. *The Powers That Be: Global Energy for the Twenty-First Century and Beyond*, Chinese Edition. Translated by Y. Song, et al. Beijing: China Machine Press, 2012.

National Natural Science Foundation of China. *China Subject Developing Strategy: Energy Science* (In Chinese). Beijing: Science Press, 2012.

O'Keefe, P. et al. *The Future of Energy Use*, Second Edition, Chinese Edition. Translated by Z. Yan, et al. Beijing: Petroleum Industry Press, 2011.

Organization for Economic Cooperation and Development/International Energy Agency (OECD/IEA). *Creating Markets for Energy Technologies*, OECD/IEA, Pairs, 2003.

Parker, S.P. *Encyclopedia of Energy*, Chinese Edition. Translated by H. Cheng, et al. Beijing: Science Press, 1992.

Qi, Y. *Annual Review of Low-Carbon Development in China (2011–2012)* (In Chinese). Beijing: Social Sciences Academic Press, 2011.

Qian, B. *New Energy Vehicles and New Energy Storage Battery and Thermoelectric Conversion Technology* (In Chinese). Beijing: Science Press, 2010.

REN21 Renewable Energy Policy Network. Renewable 2005: Global status report, Worldwatch Institute, Washington, DC, 2005.

Richter, B. *Beyond Smoke and Mirrors: Climate Change and Energy in the 21st Century*, Chinese Edition. Translated by Z. Yan. Beijing: Petroleum Industry Press, 2011.

Rifkin, J. *The Third Industrial Revolution: How Lateral Power is Transforming Energy, the Economy, and the World*, Chinese Edition. Translated by T. Zhang, et al. Beijing: China CITIC Press, 2012.

Stavrianos, L. *A Global History: From Prehistory to the 21st Century*, Seventh Edition, Chinese Edition. Beijing: Peking University Press, 2006.

Tang, F. *New Energy War* (In Chinese). Beijing: China Commercial Publishing House, 2008.

Tu, Z. *The Big Data Revolution* (In Chinese). Guilin: Guangxi Normal University Press, 2012.

United Nations Economic Commission for Europe (UNECE). Dwellings by period of construction, Chapter 6 in *Annual Bulletin of Housing and Building Statistics for Europe and North America*, UNECE, Geneva, http://www.unece.org, 2000.

United Nations. Framework Convention on Climate Change. http://unfccc.int/2860.php, 1992.

United States Environmental Protection Agency (EPA). Air emissions trends, EPA, Washington, DC, http://www.epa.gov/airtrends/2005/econnemissions.Html.

van Loon, H.W. *Ancient Man: The Beginning of Civilizations*, Chinese Edition. Translated by Y. Wang, et al. Nanjing: Jiangsu Phoenix Literature and Art Publishing House, 2012.

van Santen, R. et al. 2030: *Technology That Will Change the World*, Chinese Edition. Translated by J. Liu, et al. Beijing: China Commercial Publishing House, 2011.

Wang, G. *Introduction to New Energy* (In Chinese). Beijing: Chemical Industry Press, 2008.

Wang, Y. et al. *Smart Energy* (In Chinese). Beijing: Tsinghua University Press, 2012.

Watson, R. *Future Files: A Brief History of the Next 50 Years*, Chinese Edition. Translated by Q. Zhang. Beijing: Jinghua Press, 2008.

Wu, G. *Science Course*, Second Edition (In Chinese). Beijing: Peking University Press, 2002.

Wurfel, P. *Physics of Solar Cells: From Principles to New Concepts*, Chinese Edition. Beijing: Chemical Industry Press, 2009.

Xing, Y. et al. *Modern Energy and Power Generation Technology* (In Chinese). Xi'an: Xidian University Press, 2012.

Zhang, G. *Report on China's Energy Development for 2009* (In Chinese). Beijing: Economic Science Press, 2009.

Zhang, N. et al. *Central Asia Energy and Big-Power Politics* (In Chinese). Changchun: Changchun Publishing House, 2009.

Zhao, Z. *A Brief History of Creation* (In Chinese). Beijing: Peking University Press, 2010.

Zhu, J.C. *Autonomy and Harmony* (In Chinese). Wuhan: Wuhan University Press, 2012.

Zhu, Z. *The Future of Human* (In Chinese). Shenyang: Liaoning Science and Technology Press, 2010.

Index

Note: Page numbers followed by *f* indicate figures; those followed by *t* indicate tables.